Philadelphia's First Fuel Crisis

This volume was published with the cooperation and support of the Pennsylvania Historical and Museum Commission and the Wyoming Historical and Geological Society (Wilkes-Barre) in their continuing efforts to preserve the history of the people of the Commonwealth.

Philadelphia's First Fuel Crisis

Jacob Cist and the Developing Market for Pennsylvania Anthracite

H. Benjamin Powell

The Pennsylvania State University Press
University Park and London

Library of Congress Cataloging in Publication Data

Powell, Howard Benjamin, 1937–
Philadelphia's first fuel crisis.

Includes bibliographical references and index.
1. Anthracite coal—Pennsylvania—History.
2. Cist, Jacob, 1782–1825. 3. Businessmen—
United States—Biography. I. Title.
HD9547.P4P63 338.2′ 7′ 2509748 77-88471
ISBN 0-271-00533-5

Designed by Glenn Ruby

Printed in the United States of America

For my wife Edna and our son Stephen

Acknowledgments

During the last decade while doing the research and writing for this volume, I became indebted to more individuals and institutions than can be mentioned in this brief acknowledgment. The directors and staff of the Pennsylvania Historical and Museum Commission as well as the Wyoming Historical and Geological Society deserve special mention for helping to finance the project and for encouraging my work. I am particularly grateful to Ralph L. Hazeltine, director emeritus of the Wyoming Society, who shared my opinion that Jacob Cist deserved a business biography placed in the broad context of the developing anthracite trade. Like the Wyoming Society, the Academy of Natural Sciences has granted me special privileges to use Cist's drawings from their manuscript collections. My colleagues at Bloomsburg State College have also helped in many ways for which I am grateful. I also want to thank the institutions mentioned in the notes for their gracious assistance.

Several friends have read this manuscript and have offered helpful suggestions. They include Jacob E. Cooke, Ralph L. Hazeltine, John N. Hoffman, James P. Rodechko, and William Siener. The staff of The Pennsylvania State University Press has also been especially helpful in preparing the manuscript for publication. I accept sole responsibility for any mistakes of fact or interpretation which may appear.

Finally, and of greatest importance, I want to thank my wife Edna, who has served as a research assistant, typist, proofreader, and close companion throughout the project. In dedicating this book to her and our newborn son, I express my deep appreciation.

Contents

Figures

Introduction

> In this country every suggestion that brings forward the importance of coal to the public view is of moment: we know little of its value in Pennsylvania as yet. All, all the superior wealth, power and energy of Great Britain, is founded on her coal mining. Coal mining produce manufacturing; manufacturing produce canals, turnpikes roads and railways; these produce communication of interest, knowledge, and improvement all over the country.
>
> Where coal mines are, there, and there only will manufacturing flourish. Manufacturing produce division of labor, corporal and mental energy; they vanish idleness; they produce perpetual demand for and bring into play chemical science, and the science of machinery. They have produced a class of men almost peculiar to England, the civil engineers, to whom the country has been more indebted than to any other class of citizens, dependent on their talents for their living. All rest upon coal.[1]

So stated Thomas Cooper, one of America's leading scholars, who served as a professor of chemistry and mineralogy at several leading American colleges, sometime during the first two decades of the nineteenth century.[2] How did the belief expressed by Cooper become reality?

The problem of bringing coal to public attention was essentially one of marketing. In the 1790s Tench Coxe made a survey of the products possessed by the United States which noted the location of coal in Virginia and other places. Some of the coal deposits which could be tapped by the four leading Atlantic coastal cities lay in the anthracite region of northeastern Pennsylvania.[3] If Pennsylvania anthracite was to become an important product in the American economy, three groups had to become interested in its use. Businessmen had to decide whether the profitability of investing in mining and costly transportation projects was worthwhile or if they should support legislative action for the state to undertake the work. Merchants had to realize that this product was worth handling. And ultimately consumers had to be convinced that Pennsylvania anthracite possessed qualities superior to any other type of coal. In many cases these groups asked scientists to help them make up their minds about the feasibility of a project or the quality of a product. By studying the ways in which a market was established for Pennsylvania anthracite a modern businessman can better understand the historical dimensions of marketing and can compare the problems of the past with those of the contemporary world.

Prior to 1820 little did businessmen realize that during the next two decades the northeastern United States would turn from wood to Pennsylvania anthracite as its chief fuel. According to Alfred D. Chandler, Jr., a modern economic historian, Pennsylvania anthracite caused the origins of the Industrial Revolution in the United States.[4] From 1830 to the 1870s Pennsylvania produced more anthracite than bituminous coal. During that forty-year period, whenever a person in one of the Atlantic seaboard cities mentioned coal, he usually meant Pennsylvania hard coal.

How did early businessmen establish a market for Pennsylvania anthracite? A few popular as well as scholarly books have touched on this topic. To date, however, no one has examined this question in sufficient detail, primarily because the manuscript sources for businessmen such as Necho Allen, Erskine Hazard, George F. A. Hauto, George Shoemaker, Josiah White, and the Wurts brothers, to mention a few prominent pioneers of the anthracite trade, have been unavailable for study.[5] Only the papers belonging to the family of Jacob Cist give a clear picture of what transpired during the early pioneering days.

The forces which transformed a country dependent upon firewood to an anthracite-consuming economy must be examined in light of the elementary concepts of supply and demand. The need for coal in cities along the Atlantic seaboard had to reach a certain point before costly transportation projects could be undertaken. A severe coal shortage occurring in Baltimore, Philadelphia, and New York during the War of 1812 hastened the day when businessmen made important decisions to develop the anthracite trade.

While the fuel crisis lasted only two years, it sparked the imaginations of many businessmen who thought in terms of long-range economic development. The peace which came in 1815 caused American merchants to look for opportunities to reinvest their fortunes made during the Napoleonic wars. The efforts of New Englanders to invest in the shoe and the textile industries and of New Yorkers to innovate in auction sales, packet lines, the Latin American trade, and the Erie Canal have been documented.[6] To a few historians, merchants in Philadelphia and the Keystone State appeared very conservative from 1812 to 1825 when compared to their northern neighbors. They merely reacted to crises such as that created by the Erie Canal, and they proved generally reactionary in solving their problems.[7] However, economic historians have missed the fact that between 1815 and 1834 nearly $10 million was committed to construct canals from the anthracite deposits of northeastern Pennsylvania to the Atlantic seaboard, thereby providing the way for a cheap fuel to reach the expanding American market. More than three-quarters of this sum came from private sources at a time when it was necessary for New York State to underwrite the entire cost of the Erie Canal. The Delaware and Hudson Canal, the Lehigh Coal and Navigation Company, the Schuylkill Navigation Company, and the Union Canal each cost more than $1.5 million; and most of the money for these projects came from the mercantile community. Furthermore, Pennsylvania paid over $3 million to construct the Delaware Division Canal and the North Branch Canal for the specific purpose of hauling anthracite.

Among the merchants who strongly advocated the use of Pennsylvania anthracite was Jacob Cist of Wilkes-Barre. Cist's papers include Thomas Cooper's previous statement about coal. One of the dominating passions of Cist's career dealt with the task of bringing Pennsylvania anthracite to public notice. Consequently, his work provides a good case study in the entrepreneurship of the anthracite trade.

Definitions of entrepreneurship differ among scholars in the field. According to a leading economist, entrepreneurship is creative response: an entrepreneur is the active agent of change who makes decisions affecting a large part of the population. He functions primarily as a profit-seeking innovator who changes the nature of a segment of the economy and is the most important factor in the economy accounting for such change.[8] This definition will serve as the major concept for studying Jacob Cist's contributions in establishing a market for Pennsylvania anthracite.

However, other concepts concerning entrepreneurs must be presented in this study. The theoretical literature dealing with entrepreneurship is fairly extensive, and it is not the purpose here to discuss these concepts in depth.[9] But the reader should be aware of them and regard them as useful tools in examining Cist's contributions. As I have previously noted, the leading concept is the fact that an entrepreneur is an innovator—one who causes economic change to occur. Second, historians have borrowed the sociological concept of role-playing as a major causal factor in examining the behavior of businessmen.[10] Third, economists have created the concept of leading sector to explain the phenomenon of economic growth. According to this theory, one area of the economy starts growing faster and stimulates the growth rate of other economic activity. Fourth, the entrepreneur must be viewed in the context of the marketplace with the forces of supply and demand operating to establish prices and profits which directly affect human activity. Finally, the entrepreneur must be seen in the broad context of society where the forces of family relationships and those of politics and historical events are at work. Considering this last point, the entrepreneur cannot be strictly confined to economic theories and quantification.

A study of Jacob Cist's life raises a number of questions about entrepreneurial history and provides historical evidence to answer questions asked by scholars in the field. What role did Cist play in initiating the anthracite trade, and what were his motives? Did Cist have any special advantages which others did not possess in placing him at the forefront of the movement? How typical an entrepreneur was Cist when compared to other pioneers of the anthracite trade as well as the average early nineteenth-century businessman? This case study provides historical evidence that answers these questions and contributes to a better understanding of nineteenth-century entrepreneurial and business history.

1 Family Enterprise

Jacob Cist played an important role in establishing the anthracite trade. In order to understand his activity, the reader must view his place in the broader context of family relationships. In an article titled "The Entrepreneur and the Community," Robert K. Lamb points out the importance of the extended kinship family upon the economic growth of an industry.[1] His concept is equally valid for understanding the role of Jacob Cist because without the vision and training received from his father as well as the support of his father-in-law, uncle, and others, Cist never would have achieved his objective.

On 13 March 1782, Jacob Cist was born in Philadelphia to Charles and Mary Weiss Cist. His given name most likely came from his uncle, Jacob Weiss, who along with his father exerted a strong influence on his career. As the oldest son in a large family, Jacob assumed a position of responsibility and leadership at an early age. He grew up in America's leading port, which also served as the nation's capital for most of the years that he lived there.

Jacob's father occupied a position of prominence in Philadelphia during the years following the American Revolution. Charles Cist came from a German mercantile family. He had studied medicine briefly at the University of Halle and arrived in Philadelphia in 1773. He found employment in Philadelphia as a journalist and publisher, a career he pursued until his death in 1805.[2] Cist's profession placed him in the center of city life and made him a community leader. His publications revealed a deep concern for Philadelphia's urban problems. As a journalist and community leader, Cist's attention turned to the subject of coal when the city faced the problem of getting fuel.

In 1787 Charles Cist helped establish the *Columbian* magazine as a partner in the short-lived venture. The *Columbian* featured articles about economic activity. Among the essays aimed at encouraging the growth of American industry appeared a number on British coal. In 1788 the magazine published an article on James Brindley, one of England's foremost civil engineers, describing the coal regions of Great Britain. Brindley engineered the famous Bridgewater Canal, thereby initiating an English coal craze in 1769. He went on to construct several other coal canals which helped stimulate the British Industrial Revolution.[3] The article contained information essential to the understanding of coal production. Implicitly, it suggested that the United States undertake canals to open its coal deposits.

The *Columbian* also contained articles about coal consumption. For homeowners who were spending much money for fuel, one article advocated the use of stoves instead of fireplaces. Stoves burned less fuel and helped keep living costs down. Another article described Henry Cort's method of converting pig iron into malleable iron. Cort's air furnace consumed coal in the process of revolutionizing iron production.[4] Tench Coxe, Charles Cist's contemporary and a leading exponent of American industry, recommended adopting the latest British techniques and inventions in order to promote American prosperity. The *Columbian* favored such ideas.

By praising men for constructing coal canals and by describing inventions for coal consumption, the *Columbian* helped promote a market for the use of coal. Political leaders felt that only when the new nation gained a substantial measure of economic self-sufficiency could the United States maintain its political system. Through his magazine Cist participated in the drive for greater economic freedom. To him and others like him, development of the American coal trade was vital to the new social and economic scheme.

Just how much Jacob Cist was impressed by his father's journalistic activities and involvement in Philadelphia affairs is difficult to determine. The urban leaders and businessmen who frequented his father's shop in the 1780s certainly left an impression. When it came time to further his education after he was graduated from elementary school, Jacob chose to enter an academy which emphasized the training of businessmen.

In September 1794 Charles Cist sent his 12-year-old son to the Moravian Academy in Nazareth, Pennsylvania, where he remained for three years. Jacob was sent to Nazareth because the school had an excellent academic reputation and admitted students who wanted to pursue business careers. Religion also may have entered the picture since the Cist family belonged to the Moravian church. Furthermore, two of Jacob's cousins, Francis Weiss and Jacob Weiss, Jr., had been educated there.[5] From them, of course, he learned more about the institution. The Cist and Weiss families maintained close personal ties, and this relationship continued throughout Jacob's life.

A number of subjects in which Cist received instruction were useful in the coal business. Every merchant needed to know mathematics. Lessons in writing, English, and foreign languages developed skills used by the youth. In class Jacob may have copied sections from the standard geography textbook of that period, which located coal at Wilkes-Barre and in western Pennsylvania. Under the instruction of an old French drawing master, he became proficient in sketching natural scenery and other objects.[6] A picture drawn by Jacob in 1795 shows a factory scene with a gristmill and several other types of shops. Even at the age of 13, Cist paid attention to American manufacturing. At that time, several Philadelphia merchants had already begun investing a considerable sum of money to develop their city's industry.[7]

Moving north from Philadelphia to Nazareth brought Jacob Cist into closer contact with one of his father's business ventures—the Lehigh Coal Mine Company. As Jacob later told his father-in-law, he carefully studied the activity of the Lehigh Coal Mine Company and took notes concerning its operation.[8] He learned to know the stockholders in the company and occasionally may have served as his father's liaison with those partners living in Nazareth as well as his uncle, Jacob Weiss.

In 1785 Jacob Weiss settled near Fort Allen on the upper Lehigh Valley at present-day Weissport and engaged in the lumber business. Even though timber covered his large tract of land, a small quantity of anthracite could be used by his blacksmiths, who recognized it as superior fuel for doing their work. A Wilkes-Barre tavern keeper said that blacksmiths preferred to use hard coal because it produced greater heat than charcoal. The very high temperature melted iron faster and consequently saved valuable time. He believed that a blacksmith could do one-third more work in a day with anthracite than with charcoal. Therefore, time spent by Weiss's men in going a distance of 15 or 20 miles to nearby coal outcrops would be more than compensated for by the extra output of horseshoes and nails. Weiss must have accepted these facts. In November 1785 he sent two employees to dig coal, and records indicate that in 1787 another man spent three days digging coal for him.[9] Weiss was one of the earliest individuals in the Lehigh Valley to start using anthracite.

The exact location of the mine and the amount acquired from it remain unknown. A road went from the upper Lehigh across Broad Mountain through Beaver Meadows to Nescopeck Creek.[10] Travelers walked such paths, looking for signs of mineral wealth. Weiss himself may have numbered among those who had mastered the methods of hunting for subterranean treasures. Black meadowland and slaty soil indicated the existence of coal. Explorers examined the roots of uprooted trees searching for particles of coal. They followed barren ridges seeking exposed outcrops.[11] Each of these signs marked the existence of coal at present-day Beaver Meadows. A 1783 report on mines in the United States said that firm, shining, loose pieces of coal had been found on top of Broad Mountain.[12] This evidence suggests the mine was at Beaver Meadows. The road allowed coal to be hauled back to Fort Allen.

Jacob Weiss numbered among the earliest coal prospectors. As soon as the American Revolution ended, at least two European geologists walked through the woods of northeastern Pennsylvania hunting for anthracite. In their accounts, each described natural signs associated with coal formations. A German natural scientist said in 1783 that people in the Lehigh Valley continuously dreamed of mines and sudden wealth.[13] In 1784, the publication of a 37-page manual describing methods for finding mineral wealth fed such hunger. A member of the Weiss family copied just those sections of the pamphlet dealing with coal.[14] Since Jacob Weiss knew about one coal mine on the upper Lehigh, chances existed that with guidance he could find others.

The manual listed five signs for coal, which may have proved valuable for locating mines in the Lehigh Valley. It stated that rocky barren mountains had the greatest probability of containing rich mines and recommended following the top of barren ridges and inspecting the color of rocks. Black slaty rocks indicated coal. Prospectors could detect coal from the color of grass and the smell of vapors. If springwater contained black sediment, coal might exist nearby.[15]

By coincidence, if not directly, the barren mountain principle bore fruit in the discovery of the famous Lehigh Coal Mine. In 1791, Philip Ginther came across an immense outcrop of anthracite at present-day Summit Hill. The day after his discovery Ginther carried the anthracite nearly 20 miles to Jacob Weiss at Fort Allen. Ginther's actions indicated, therefore, that he knew the methods for detecting coal and was eager to verify his discovery.

His samples met with Weiss's approval. Then Weiss took the news and samples to his brother-in-law, Charles Cist, and other leading Philadelphia financiers. Following Quaker tradition, these men showed a willingness to invest in mining ventures.[16]

Other reasons were equally important. In 1790 Philadelphia had a population of 42,000, and the location of the national capital in the Quaker City at that time insured continued growth. Even at this early date, Philadelphia had begun to depend on coal. Although the statistics are imperfect, they indicate that anthracite and bituminous coal had frequently come into the city as ship ballast throughout the colonial period. The nonimportation agreements signed by Philadelphia merchants to deal with the Stamp Act in 1765 enumerated coal as one of the exempt products.[17] Craftsmen wanted coal to ply their trade.

As the number of craftsmen in Philadelphia increased after the

Revolution, the demand for coal grew. In 1784 the Quaker City imported just over 500 tons from places such as London, Liverpool, Glasgow, and Virginia. The customs records also show that from 1790 to 1792, the average amount of coal entering Philadelphia exceeded 1,000 tons annually. Approximately one-third of this quantity came from Ireland, which produced a type of anthracite closely resembling that of the Lehigh Coal Mine.[18] Although the number of artisans was small, they made up a significant segment of Philadelphia's economy.

To Jacob Weiss, Charles Cist, and other merchants, the opportunity to develop a good coal mine near Philadelphia that would supply craftsmen as well as a growing urban population appeared particularly attractive. If sufficient capital could be secured to develop the Lehigh Coal Mine and market its coal in the city, the merchants hoped to establish a profitable business. Knowing that John Nicholson, a prominent Philadelphia politician and mercantile speculator, was establishing an industrial town at the Falls of the Schuylkill where anthracite could be used by craftsmen and burned by steam engines, Jacob Weiss and Charles Cist decided to take the Lehigh anthracite to him.[19] The diary of Nicholson, who was also comptroller general of Pennsylvania, reveals a strong initial enthusiasm for the project. On 19 January 1792 the group gathered to discuss plans. One week later Nicholson and Weiss met to draw up a prospectus that would be circulated among friends. Before the group issued a public statement on 13 February calling for the organization of the Lehigh Coal Mine Company, Nicholson spent an afternoon answering questions asked by prospective stockholders.[20]

The businessmen wanted to form an unincorporated joint stock company worth $10,000. Jacob Weiss received 10 of the 50 shares, and $8,000 for his services, the coal mine, and 770 acres. As soon as financiers subscribed the remaining shares, the company would organize, elect officers, and assess each share $10 to launch the enterprise. The businessmen felt that with $500 they would be able to commence mining, construct a crude road, and send some anthracite to market. Stockholders received the power to levy additional sums as long as the majority agreed. The plan also stipulated that if the mine did not contain sufficient coal to last 10 years, subscription money would be returned to the investors.[21]

Promoters of the Lehigh Coal Mine Company found a ready market among financiers for shares of their company stock. Within one week every share was subscribed with the distribution falling into the hands of two groups. The Cist-Weiss family and their friends made up the first group. In addition to Jacob Weiss's 10 shares, four of his friends who attended the Moravian church at Nazareth purchased 10 shares.[22] Consequently, Nazareth and Fort Allen became the two hinterland outposts for developing the coal mine. In Philadelphia Charles Cist picked up four shares and another merchant and friend of the Cist family acquired six, thus giving this group 60 percent control of the company. The remaining stock went to John Nicholson and his associates. Nicholson kept eight shares and gave two shares to a friend. Nicholson induced still others to become partners in the venture. He wrote to a prospective investor that the coal mine and land were worth the price, and the business would earn its owners good profit.[23]

At the first corporate meeting on 21 February 1792, the stockholders elected John Nicholson president.[24] Respect for Nicholson ran high among the group. As comptroller general of Pennsylvania and as a reputable Quaker merchant, he had the right political and economic connections which, no doubt, would help the new enterprise. During 1792 he spent several hours each week handling company affairs. His diary reveals careful attention to business details. The president and his managers met regularly the second Tuesday of every month to discuss the mine. Of the 17 organizations to which Nicholson belonged, he gave as much time and attention to the Lehigh Coal Mine Company as he gave to any other group. Although he became more interested in land speculation during the next few years until he went bankrupt, Nicholson remained company president until 1798.[25]

One policy which Nicholson pursued enlarged the initial coal company acreage. In 1792 the company embarked on an extensive land acquisition program. Nicholson put up money for land office fees and warrants. After receiving warrants for unclaimed land in the upper Lehigh Valley, Jacob Weiss employed men, purchased supplies, and surveyed tracts averaging 400 acres apiece. He located the land, marked off boundaries, and sent the surveys back to Philadelphia. Then Nicholson paid for the patents and secured deeds. The company assessed each share of stock an equal amount for the expense of acquiring land. By 1798 the company had secured a total of 8,665 acres.[26] The land covered an area approximately 4 miles wide by 14 miles long stretching from the Lehigh River at Mauch Chunk (now Jim Thorpe) on the east to the Little Schuylkill River at Tamaqua on the west. The physical dimensions of the estate remained virtually unchanged for the next 150 years. The land became the coal property of the Lehigh Coal and Navigation Company, which included the entire eastern end of the southern anthracite field.

To assume responsibility for mining coal, Nicholson secured the services of his friend, Thomas Bedwell, in Philadelphia. Apparently, Bedwell was a self-educated mining expert who carefully copied into a notebook a 19-page illustrated encyclopedia article concerning the British coal industry. The article covered such major points as prospecting methods, geology, mine layout and technology, mining technique, and transportation. This notebook later became the property of Jacob Cist, who was interested in learning about the British coal industry.[27]

During 1792 the project developed rapidly. On 17 May a report favored the Summit Hill mine which contained an almost inexhaustible coal bed and signs of nearby iron ore deposits. The committee recommended that a nine-mile road to the river be constructed and urged managers to take steps in order to put the coal mine into production. The managers accepted the report and immediately voted to assess each share $5 to cover costs for opening the mine.[28]

That summer, workmen cleared debris off the mine face, started digging, and carted coal to the river bank. By the end of August some coal had reached the river and soon descended the Lehigh and Delaware rivers to Philadelphia. A sample of Wyoming Valley coal also went down the Lehigh River to the Quaker City. Mining coal continued into the fall, and on 21 September Nicholson gave the workmen money to dig more anthracite.

Nevertheless, the initial enthusiasm of the group faded after one year. While Jacob Weiss took over the mining responsibilities for the rest of the decade, Charles Cist handled most of the marketing responsibilities in Philadelphia. Jacob Cist, no doubt, helped his father with some of this business after he was graduated from Nazareth Academy in 1797. For the next three years Jacob worked for his father in Philadelphia.

Since Fort Allen was only a short distance from the mine, Jacob Weiss could secure men to dig and haul coal for several weeks a year. On 9 August 1793 company stockholders passed a resolution calling for mining as much coal as demand permitted. Evidence shows that coal came from their mine each year. A Northampton blacksmith testified that he acquired anthracite annually from 1794 to 1814. On 14 July 1801 Charles Cist wrote to Jacob Weiss's son, Francis, that a fresh supply of coal that season would be welcome.[29] While output stayed small, demand for anthracite in the Lehigh Valley and Philadelphia kept the mine in production.

A major limitation to increased output was the high cost for hauling anthracite nine miles from Summit Hill to the Lehigh River. The expense of sending coal by wagon kept market prices high compared with coal coming from Great Britain and Virginia. From the start management recognized the high cost of land transportation. Desiring to lower operating costs and to expand sales, they searched for mines closer to the Lehigh River.[30]

In their search for coal near the river, the Lehigh Coal Mine Company turned to European techniques. In the notebook Thomas Bedwell made, there was a section on the methods of finding coal seams. According to the notebook, coal seams intermixed with other kinds of strata descended at different angles from the surface to various depths. Strata generally lay with uniform regularity underneath hills and valleys. When looking for coal, a person should examine all places where natural conditions have exposed solid strata to view. In a memorandum book he should note the different thicknesses of strata, the order in which they lie on each other, the front of the horizon to which they dip or incline, and whether they lie in a regular manner. This should be done in every place on the terrain where solid strata can be carefully inspected. Then one can compare the outcrop in all exposed places as to kind of substance, thickness, and incline. If uniformity of strata can be seen in a number of places, then a fairly accurate outline of a formation can be drawn. In places where coal strata have remained unexposed, the notes recommend boring holes at least 60 feet deep and measuring the depths of different strata. By drilling holes, a person can determine the depth of coal veins, their thickness, and angle of inclination.[31] This information eventually proved useful to Jacob Cist even though it helped the mining company very little during the 1790s.

The company's search for coal close to the river proved unsuccessful. Only the mine at Summit Hill produced coal. During the 1790s workers quarried coal from open pits. In 1804 the three or four small pits resembled crude wells into which a person could descend with ease and acquire coal. By that date, one observer believed, the mine had yielded several hundred tons.[32]

The Lehigh company also faced the problem of getting coal to market. Wagons carried coal to consumers living in Northampton County communities such as Fort Allen or Nazareth. For those who desired anthracite, a wagon trip of 25 or 30 miles overland fell within normal expectation. Residents took similar trips to county seats and neighboring towns to transact business. Those who wanted hard coal spent the required time and energy to get it. Wagons also carted Lehigh coal 100 miles from the mine to Philadelphia, especially if they happened to be returning empty after taking merchandise from the Quaker City to stores at Fort Allen or Wilkes-Barre.

Obviously, shipping coal down the Lehigh River on a raft to Allentown, Bethlehem, Easton, or Philadelphia was the cheapest way of getting anthracite to market. Prior to the American Revolution, Pennsylvania declared the Lehigh River a public highway. Starting in 1785 Jacob Weiss used the Lehigh River to send grain and lumber downstream to market. Marks placed on large rocks along the riverbank designated the proper water depth for rafting. As soon as the water level rose to these marks, raftsmen shoved off for the swift ride down river. They learned every impediment and rapid in the river and developed skills enabling them to guide large rafts safely down dangerous currents. Employment depended upon one's ability to get goods undamaged to market. By 1790 a corps of skilled raftsmen made a living by boating. During the spring and fall, they took country produce to market. Their talent had become essential for conducting the coal trade.

Merchants such as Jacob Weiss recognized the necessity of removing as many impediments to river traffic as possible. They continuously petitioned the state legislature for funds to accomplish this task. In 1791 Pennsylvania appropriated £1,000 to improve navigation of the Lehigh.[33] At one of the earliest meetings of the mining company, the managers created a fund for removing some of the treacherous obstacles that hindered boating. A Nazareth stockholder accepted the assignment of examining the river from Mauch Chunk to Bethlehem and calculating costs which would improve navigation for boats carrying 10 or 12 tons. He reported that £7,000 would be needed to clear the Lehigh so that anthracite could go downstream during dry season.[34] In spite of his recommendations, limited funds precluded any extensive improvement. Coal waited on the river bank until high water permitted shipment. This problem would plague Jacob Cist and his partners during the War of 1812.

The Lehigh Coal Mine Company took steps to encourage hard coal consumption. In February 1794 the managers ordered a Nazareth stockholder to transport coal to Allentown, Bethlehem, Easton, and elsewhere so that it would come into more common use. The stockholder received authority to give five bushels to anyone who was willing to experiment with anthracite. In 1800 Francis Weiss carried on his work. He wrote and talked to numerous craftsmen in the Lehigh Valley about burning hard coal. In Philadelphia Charles Cist worked at getting Lehigh anthracite accepted. He also kept in contact with Weiss about coal-burning experiments.[35] The early adoption of Pennsylvania anthracite depended on the efforts of these entrepreneurs.

Lehigh anthracite gradually gained acceptance. At first the new fuel met with mixed reaction. However, overall testimonials endorsing its value outweighed negative reports made by unsuccessful experimenters. By the end of the eighteenth century, blacksmiths living in the Lehigh Valley generally accepted hard coal as superior fuel for doing business. They discovered the same qualities in the fuel which Jesse Fell had spoken about earlier. The hotter temperature and the time saved made hard coal a favorite fuel. A Bethlehem blacksmith told Francis Weiss he preferred Lehigh coal to bituminous and charcoal. He would rather pay $.50 a bushel for anthracite than $.07 a bushel for charcoal. After learning the techniques of burning hard coal, other craftsmen generally agreed with the Bethlehem blacksmith.[36] While its relative merits varied, Lehigh Valley blacksmiths desired anthracite because it was a more efficient type of fuel.

Philadelphia started consuming Lehigh coal during the 1790s. In 1801 Charles Cist wanted several hundred bushels for his own use. He also needed anthracite for trials at the glass house and steam engine. He calculated that the new steam engine for the city's waterworks would consume tons of anthracite annually. In 1794 Tench Coxe said that a wide variety of industry depending on fire for producing goods burned coal. Cist took samples to a number of Philadelphia industries so that they could try Lehigh anthracite. For some the trials failed, and they threw the coal away. Others liked hard coal. Cist informed Francis Weiss that several firms wished to burn anthracite on a large scale if they could be certain of receiving a regular and sufficient supply.[37] By means of personal contact with artisans, Charles Cist in Philadelphia and the Weiss family in the Lehigh Valley slowly established a market for hard coal.

Urban homeowners could also burn anthracite. Benjamin Franklin invented a fireplace and two stoves that used coal. Within a decade his famous Pennsylvania fireplace of 1741 had been converted to coal. In 1785 he published a pamphlet on stoves which described procedures for burning coal. Anthracite as well as bituminous coal would burn in his stoves.[38] Lehigh company managers experimented with their coal in stoves. In 1796, Godfrey Haga, a Lehigh company manager, asked John Nicholson for two bushels of coal to see if it would burn in an open stove. Many people found Lehigh anthracite difficult to kindle and consequently refused to purchase it.

In 1798 Oliver Evans became interested in anthracite. He studied the principles for burning the most incombustible kinds of fuel and invented a stove for Lehigh anthracite. He constructed 4 small stoves to burn hard coal. For the experiment he acquired 30 tons of coal from Nicholson. The most incombustible Lehigh coal, which would not burn in open grates, was easily kindled in his stove. The flame gave off a brightness far exceeding any open fire and created light sufficient for reading newspapers. Evans said his stoves made the most durable, clear, and agreeable fire he had ever seen. After receiving a patent in 1800, he called a meeting of the Lehigh Coal Mine Company managers and demonstrated his device.[39] Consequently, Evans became a strong advocate for the introduction of Lehigh anthracite. Slowly Philadelphia awakened to the possibility of heating homes with Lehigh anthracite.

To finance the coal company required money, and without company account books it is impossible to determine how much each stockholder paid annually. Bills for land acquisition, coal prospecting, mining, road construction, river improvement, and transportation amounted to several hundred dollars each year. As the 1790s progressed, assessments increased. In 1796 Charles Cist owed $500 on his four shares. He paid $420 and remained indebted for $80. Three other stockholders owed the company a total of $90.[40] The transportation project in 1798 called for even more money. Throughout the decade Cist and others contributed substantial sums to the company.

Like other Philadelphians, Charles Cist diversified his investments, and Jacob helped take care of several of these projects. In 1797 Cist became the official government printer of John Adams's administration. He also began investing in several projects to develop Washington, D.C. When the national capital moved from Philadelphia to Washington in 1800, Jacob received a clerkship in the post office. He managed to retain the position during Thomas Jefferson's administration and looked after his father's business affairs in the national capital until he moved to Wilkes-Barre. Charles Cist's business suffered from the change in administrations, and this experience may have taught Jacob a lesson. For the rest of his life he remained nonpolitical even though he served as the postmaster of Wilkes-Barre from 1808 to 1825.[41]

Charles Cist died in Lehighton in 1805 while searching for coal land. Among his possessions were four original shares in the Lehigh Coal Mine Company. Two years later Jacob considered selling the stock. He felt that the company managers acted rashly by incurring many expenses which should have been postponed until they were absolutely certain of success. He foresaw misunderstandings and lack of union which attended "those many headed monsters" called companies. By March 1808 he had made a conditional agreement with an Englishman for the four shares belonging to his father's estate. But the deal never materialized. Six years later Cist still retained the shares which eventually enabled him to help initiate the Lehigh coal trade.[42]

Jacob Cist's opportunity to become the leading advocate for the development of the Pennsylvania anthracite trade arose in 1807 when he married the daughter of the richest merchant in northeastern Pennsylvania. On 29 October 1807, Cist informed his father-in-law, Matthias Hollenback, a Wilkes-Barre resident, that he would accept a partnership in the family business. Hollenback's business extended from Philadelphia and Baltimore up the Delaware and Susquehanna valleys and westward through lower New York State to Buffalo. For a young gentleman this was the opportunity of a lifetime and a chance to provide a good living for his 18-year-old bride.

Matthias Hollenback offered to put all the merchandise and equipment at his four stores and warehouse into the partnership if Cist would contribute the net proceeds of 1,700 Spanish dollars he received from a Calcutta adventure. The partnership would run five years. During the first two years Hollenback agreed to teach Cist about purchasing merchandise, driving cattle, and arking goods down the Susquehanna River. He wanted his son-in-law to attend to the entire operation and assume responsibility for all contracts, obligations, and endorsements. For his services Cist would receive one-third of the profits and Hollenback would retain the remaining two-thirds.[43] Thus Cist was given an opportunity to make his fortune.

He also became heir to Matthias Hollenback's real estate. Hollenback was the largest landlord in Luzerne County, which at that time included present-day Lackawanna County. During the great land grab of the 1790s Hollenback ranked among the sharpest land jobbers in the Commonwealth, laying claim to over two-thirds of the Lackawanna Valley. In the Hazleton region of lower Luzerne County, Hollenback had also entered the race for land. In 1792, with a Nazareth partner, he laid claim to over 10,000 acres but sold 9,000 to John Nicholson and Robert Morris. His partner kept the acreage, which eventually became the Hazleton Coal Company estate. Tench Coxe, who also acquired several thousand acres in this area, noted as early as 1790 that anthracite existed on the head waters of the Lehigh River and Nescopeck Creek. The final district in which Matthias Hollenback secured coal land was the Wyoming Valley. Travelers visiting Wilkes-Barre noted many coal outcrops in the plains and mountains forming the valley. Hollenback became the valley's foremost landlord, holding about 3,500 acres on the south side of the Susquehanna.[44] In all, Hollenback controlled 48,500 acres of coal land. His acreage matched that which was later accumulated by anthracite railroads and made him one of the earliest anthracite coal land barons.

At least one coal mine on Hollenback's land produced coal. He owned a mine one mile east of Wilkes-Barre at the mouth of Mill Creek. Coal coming from this mine as early as 1776 was used to make guns at the Carlisle Arsenal. Four years later two more shipments of anthracite from the same mine descended the Susquehanna for Carlisle to be used in making muskets. Blacksmiths who had burned anthracite in Wilkes-Barre since 1769 could get their fuel from Hollenback. According to Jacob Cist's letters, in 1807 Mill Creek coal was used to make steel, glass, and a variety of other products.[45]

Jacob Cist recognized the opportunity of going to Wilkes-Barre to work for his father-in-law. He felt that Wilkes-Barre presented various possibilities which, if properly pursued, could not fail to make him independently wealthy. While engaged in Hollenback's business, Cist could make preparations for putting other plans into operation. His thinking reflected contemporary national trends for developing extractive industries and manufacturing. Anthracite, of course, played a major role in his planning.

Cist wanted to purchase a coal mine near Wilkes-Barre which he and Matthias had viewed together. On 13 November 1807 he informed Matthias that he must possess that coal bed and would pay $200 or $300 for it. A Philadelphia ironmaster, who needed coal to carry on his business, intended to visit Wilkes-Barre and look at coal. He planned to settle at Wilkes-Barre or near the Lehigh coal mine. Cist wanted to acquire the mine before the ironmaster's intentions became known and the owner raised the price. Five months later Jacob again conveyed to his father-in-law his anxiety about getting the coal mine. He feared that news of the ironmaster's intentions and a growing craze for coal would put the price beyond his range. From the mine's vicinity to Wilkes-Barre, Cist said he could operate a glassworks without interfering with his other duties. On that account he believed the enterprise would be very desirable.[46]

Jacob Cist also took an interest in Hollenback's Mill Creek mine. In a letter, he informed Matthias that steel could be produced with anthracite from Mill Creek. He believed that Hollenback could not hide the value of his mine long from the outside world. Several people in Washington had already asked Cist about the mine's potential, and he suggested that Hollenback show his mine to the Philadelphia ironmaster. If Hollenback's anthracite met specifications, the Philadelphian might decide to locate at Wilkes-Barre. He wanted to establish a furnace and forge as well as rolling and slitting mills. The ironmaster's enterprise would consume a large amount of anthracite.[47]

Cist considered forming a company to manufacture cannons at Wilkes-Barre. Since the federal government feared war with a European power in 1807, the opportunity existed to supply material for a national arsenal. If cannons could be taken easily from Wilkes-Barre to Baltimore, Cist felt he could get a contract from the secretary of navy. Cist suggested to Hollenback that a coal bed be attached to the cannon works. He had seen letters from an English gentleman who proposed investing several thousand pounds in such a company. Since Jacob Cist had mentioned that Hollenback owned a coal mine, he was requested to find out whether Matthias would sell it and at what price. On the other hand, Hollenback and Cist could initiate the enterprise and attract foreign investment.

A cannon foundry at Wilkes-Barre was not the only one contemplated for the northern anthracite field. In the previously mentioned letter about his plans, Cist noted that Charles Miner, a Wilkes-Barre publisher and one of his future associates who helped launch the hard coal trade, had been negotiating with the federal government to dispose of land, an ore bed, and an iron works in the Lackawanna Valley for a cannon foundry. His property contained a coal bed for fuel. Cist said the government preferred dealing with private companies rather than establishing public works. Cist felt his Wilkes-Barre project had better chances for success.

From the comments in his letters to Matthias Hollenback, it appears that Jacob Cist had well-developed ideas about the role anthracite could play in the development of Pennsylvania's economy. Many of these ideas had been expressed by his father, and they had not fallen on deaf ears. Jacob's knowledge of the Lehigh Coal Mine Company gave him an insight into the frustrations he would face. Yet he firmly believed that once a market had been established, anthracite would directly stimulate the growth of the American economy. Establishing such a market challenged his ability, and he was ready to meet the challenge.

Moving to Wilkes-Barre enabled Jacob Cist to come into contact with a number of other men who were interested in starting the anthracite trade and who later numbered among his associates. From the spring of 1808 until the fall of 1811 Cist and his young bride lived adjacent to Hollenback's Mill Creek mine.[48] At this residence he could observe and learn about the coal trade in the Wyoming Valley.

The earliest partnership to establish a permanent anthracite trade in the Wyoming Valley and down the Susquehanna River was the firm of Abijah and John Smith. They came from a Connecticut county which had supplied firewood to New York City for over a century. In 1804, a magazine article calling for the development of coal mining noted the scarcity of fuel in southern New England. The article specifically pinpointed New Haven County, the home of the Smith family, as an area destitute of firewood. Within the previous five years the price of woodland in this southern Connecticut region had risen over 200 percent. The author of the article called upon the people living in this region to consider the matter seriously before all the firewood in their area vanished. They needed to reallocate capital toward opening mines and developing the coal trade.[49] These facts were a part of the environment in which the Smith family lived and exerted an influence on their thinking.

During the same months in which Cist was writing to Hollenback about moving to the Wyoming Valley and buying a coal mine, John Smith loaded an ark full of anthracite from a mine located in Plymouth, near Wilkes-Barre, which his brother had purchased the previous year. He shipped the hard coal down the Susquehanna River to Columbia where he found someone to purchase it.

Smith's enterprise was greatly aided the following February when Jesse Fell invented a grate to burn anthracite in a fireplace. Almost every farmer in the Susquehanna Valley used a fireplace to heat his home. Other men, such as Thomas Cooper, the famous politician and scientist, had contemplated such a device, but Fell became the first person to perfect the idea.[50] For towns the size of Wilkes-Barre, which contained a population of approximately 500 when it was incorporated as a borough in 1806, anthracite offered the cheapest and best form of fuel. Since farmers also spent a disproportionate amount of time gathering firewood, they too welcomed Fell's grate and a steady supply of hard coal. Their time could be better spent pursuing more profitable commercial ventures such as raising grain to sell to the coastal cities. In 1808 Abijah and John Smith took Fell's grate down the Susquehanna on their coal boat to teach the farmers how to burn anthracite. They established coal yards at Columbia, Pennsylvania, and Havre de Grace, Maryland, and several merchants agreed to sell their coal. Their enterprise proved successful as they acquired new customers and the Susquehanna anthracite trade expanded.[51]

When Jacob Cist moved into a new home in Wilkes-Barre in 1811, he associated with a number of people who were also interested in the long-range development of the anthracite trade. Matthias's son, George, born in 1791, listened to Cist and became a convert to his cause. Charles Miner, editor of the Federalist newspaper, who, like the Smith brothers, had migrated from southern Connecticut to the Wyoming Valley, was a friend who shared Cist's views.[52] Isaac A. Chapman, another southern Connecticut migrant to Luzerne County, came under Cist's influence and became a strong advocate for launching the trade. Other key leaders who helped Cist establish a market for anthracite included David Scott, a Luzerne County judge instrumental in getting the North Branch Canal, and a Morristown, New Jersey, merchant who helped plan an anthracite railroad running from the Wyoming Valley to New York City.

Among these men, why did Jacob Cist stand out as the most important leader in launching the anthracite trade? It was not until 1812, when Cist reached age 30, that enough evidence exists to give a good view of his character. From 1815 to 1825 a clear picture emerged as Cist matured and became a leading Wilkes-Barre businessman. Evidence indicates that he practiced the essential virtues of the Protestant ethic—hard work, diligence, patience, prudence, and frugality. He worked hard attempting to introduce anthracite during the war. He put in long hours either at the store, in the field following some point of scientific inquiry, or on a business trip. Cist pursued a project diligently as indicated in the comprehensive way he approached the uses of anthracite, the systematic exploration of the hard coal formation, or the minute details in all his drawings. Diligence was also evident in the way he kept accounts, recording even the smallest matter, and in his beautiful handwriting. Unending patience also played a part in these tasks which he enjoyed. Certainly, he waited patiently for the market to develop, a dream and a movement that he saw slowly evolve.

Prudence and frugality are difficult traits to measure. There is no indication that Cist was a spendthrift or that he rushed into business affairs without thinking them through ahead of time. On the other hand, the excessive amount of time, energy, and money he put into developing the anthracite industry may have made him appear prodigal and foolish to some of his contemporaries. In retrospect, the rapid growth of the trade, starting in the 1820s, makes him look like a pioneer entrepreneur—a person with great insight and understanding who laid the groundwork and helped to establish the market.

To most of his contemporaries, Cist was a trusted and highly respected man of science and business. Since he was from a prominent Philadelphia family and had married the daughter of the richest merchant in Wilkes-Barre, Cist was in the social position for elite leadership and could expect deferential treatment. Even more important, he was a man of culture and education. He had one of the largest libraries in Wilkes-Barre, consisting of nearly 50 volumes and ranging from classics to modern works on geography, surveying, botany, and chemistry.[53] Among his contemporaries, only Isaac Chapman, Charles Miner, and David Scott appeared to have similar educational backgrounds.

Cist was also intellectually curious. He advanced scientific knowledge by studying two major natural phenomena, the anthracite formation and entomology. His drawings of Wyoming Valley insects are the subject of a special study being made by a modern student in entomology.[54] Cist's geological work will form the basis of a chapter. Detailed work in both these areas indicates that Cist was quiet, peaceful, and patient. To accomplish the type of work he did required these traits.

Cist's quiet, unassuming manner stands out in still another way. He appears as a man who worked closely with other people and instilled confidence in them. He never boasted or tried to occupy the center of the stage, yet people recognized his ability and chose him to be their leader.

He was certainly a creative thinker who understood the political and economic world in which he lived. His invention of a coal stove and ink reveal an ability to isolate the specific points of a problem, group them with other things in the environment, and finally develop something new. Being postmaster of Wilkes-Barre helped expand his horizon. As Benjamin Franklin indicated in his *Autobiography,* the postmastership brought him into greater contact with people.[55] As they arrived in Wilkes-Barre, Cist probably scanned state and national newspapers, plus other information. From these sources as well as his contact with other people in southeastern Pennsylvania and the rest of the United States, Cist learned that other businessmen were investing money and were interested in developing the hard coal trade. His vision, then, was not an isolated instance of premature thought. His calculated plans were based on substantial information about a natural resource for which there would soon be great demand.

While doing business in Wilkes-Barre and traveling through the northern coal field, Jacob Cist met many landowners and potential coal-mine operators. Many of these people paid their taxes to him during his tenure as county tax collector from 1816 to 1818. Consequently, they learned to trust and respect the young businessman. On several occasions they elected him to preside at a town meeting or to serve as recording secretary, or they put him on the correspondence committee for a special project. Cist had earned the reputation of being a sound businessman with great ability.

In addition to the aforementioned qualities, Cist had special talents which served the interests of the anthracite trade. He could write extremely well, as shown in his 1815 pamphlet on hard coal and his voluminous correspondence. Facility with words was an essential skill for promoting the trade. In letters to several leading American scientists of the day, including Thomas Cooper and Benjamin Silliman, he discussed the coal business. He was a skilled artist who drew sketches of many things, including a self-portrait, as evidenced by the illustrations in this book (see figure 1). His speech delivered at the 1825 Pennsylvania Canal Convention in Harrisburg indicates that he was a capable public speaker. By writing, drawing, and speaking, Cist articulated the desires of Wyoming Valley merchants to develop a market for Pennsylvania anthracite.

There is still another dimension of Cist's personality which deserves mention. He was very much a family man who established close personal relationships with his relatives and friends. He was candid with Isaac Chapman. In their correspondence a note of disappointment occasionally came out, but never one of hostility. Their letters reveal accurate statements of fact with an underlying tone of optimism about the coal business. This same respect and openness is shown in the letters of Philadelphia businessmen to the Wilkes-Barre merchant.

Jacob Cist remained close to his mother in Philadelphia, his kinfolk on the upper Lehigh, his relatives in Wilkes-Barre, and his other friends. On at least one occasion his immediate family determined the course of his career. In 1815 his wife and four children did not care to move to Mauch Chunk where Jacob would assume responsibility for the Lehigh Coal and Lumber Company.[56] Since they wanted to remain in Wilkes-Barre, Jacob stayed in the Wyoming Valley. Certainly this

1. Self-Portrait by Jacob Cist (Wyoming Historical and Geological Society)

decision helped endear him to the businessmen of the northern coal field, for in Cist they had a talented young man upon whom they could count.

Staying in Wilkes-Barre was not a poor choice. As a partner in the Hollenback firm, Cist shared in the profits. He did not have to remain in the store all the time but could go on business trips or scientific excursions. During this time the firm employed about ten men, with Cist, his younger brother-in-law, George M. Hollenback, and others overseeing the business. Matthias had gone into semi-retirement after 1807 when Cist assumed the greatest responsibility for the firm. Nevertheless, in his sixties and seventies, Matthias continued to be active and outlived Cist by three years.[57]

The untimely death of Jacob Cist on 25 December 1825, at age 43 brought the activities of this anthracite entrepreneur to an end. Just how much impact he had on starting the trade is impossible to determine. The range and extent of his activities placed him at the forefront of the movement. Others also took a vital interest in establishing the market, many of whom had no direct relationship with Cist. In this sense he is typical of a larger group of businessmen who from 1812 to 1825 understood the potential of anthracite and wanted to exploit this resource. In other ways Cist clearly stands at the head of the line in stating the value of anthracite to the general public, in mapping the geological formation, and in coming up with a compromise in 1825 for the Pennsylvania transportation system. The fuel crisis during the War of 1812 provided Cist and others with an opportunity to become heavily involved in starting the Pennsylvania anthracite trade.

2 Wartime Production

Philadelphia's coal crisis of 1813–1815 created an opportunity for Jacob Cist and others to become heavily involved in the anthracite trade. Cist received news of the crisis on 13 March 1813, when James Ronaldson, one of the city's leading typefounders, who used a large quantity of Virginia coal to melt lead, wrote to the Wilkes-Barre merchant that the British blockade of the Chesapeake and Delaware bays had greatly increased the price of Virginia coal. The price of Virginia coal, which had averaged about $.30 a bushel since the American Revolution, ranged from $.75 to $1.00 a bushel in March 1813, and for the next 24 months would not drop below $1.00 a bushel. Ronaldson told Cist it would be a good time to market Susquehanna coal in Philadelphia and that he might also be able to establish the trade with Baltimore.[1]

Philadelphia's fuel crisis had two dimensions. The British blockade caused a psychological as well as an economic crisis. As noted previously, by the 1790s many city craftsmen depended upon coal to carry on their trade. Ronaldson, who later established the city's leading typefoundry, Binny and Ronaldson, was only one of them.[2]

By 1813, many artisans who had preferred coal to charcoal for a number of years considered it a necessity. This fact is reflected by the establishment in 1813 of the Mutual Assistance Coal Company of Philadelphia for the Promotion of Manufactures. The coal company included craftsmen from just about every trade in the city. As soon as coal prices reached the $1.00 per bushel level, leading mechanics and industrialists such as Oliver Evans created the association to provide for regular coal shipments to Philadelphia. The company wanted to make contracts for delivering coal in the Quaker City with anyone desiring to engage in the trade. The company hoped to find coal within 20 or 30 miles of Philadelphia.[3] Failure, however, forced it to transport coal more than 100 miles from anthracite mines in northeastern Pennsylvania.

From 1790 to 1810 the number of craftsmen in Philadelphia had grown steadily as the city more than doubled its population. According to the 1810 manufacturing census, Philadelphia had 201 blacksmiths, 2 iron furnaces, 1 rolling and slitting mill, 10 gun manufacturers, 20 nail makers, 20 brick kilns, 3 lead manufacturers, 22 soap and candle makers, 17 breweries, 18 distilleries, and 3 glasshouses.[4] Every one of these businesses burned coal, some to a lesser extent than others.

Indeed coal had proved its value to Philadelphia craftsmen and leading industrialists. Oliver Evans, who in 1806 had constructed the Mars Iron Works, which used a large amount of Virginia and foreign coal, was one of the first advocates of Pennsylvania anthracite. By 1800 he had discovered the value of Lehigh coal and thereafter championed its use even though he found it difficult to acquire.[5]

Another chief consumer of coal in the Quaker City was the Philadelphia waterworks, which operated two steam engines to pump water for the growing city. On 13 November 1804 the managers of the waterworks had paid Frederick Steelwagon, one of Philadelphia's leading coal merchants, $80 on their coal bill. The waterworks used wood to run its steam engine on the Schuylkill and coal to heat the water for the center square steam engine. During a nine-month period in 1805, the center square engine consumed about 500 tons of coal. The next year wood was used instead of coal due to its lower price. At the end of 1806 the center square engine again turned to Virginia coal for its fuel, using over 500 tons from November 1806 to October 1807. The next year this steam engine consumed more than 700 tons. Upon inquiry by the American Philosophical Society in 1809, the managers of the waterworks reported that they used 1–1.25 tons of the best Virginia coal each day.[6]

As the price of firewood increased, steam engine operators turned to coal for their fuel. Federal statistics show the amount of coal that entered all American cities after 1790. While these statistics are not completely reliable, they do indicate the overall trend. During the 1790s the amount of coal entering all American ports averaged above 7,000 tons annually. During the next decade the amount increased significantly, the exception being the years during the embargo. The tonnage for this decade averaged nearly twice that of the previous ten years. In 1810 and 1811 more than 13,000 tons entered American ports each year. With the coming of the War of 1812, the total tonnage dropped off drastically to under 3,000 tons a year, with only 691 tons entering in 1814. A slight recovery over the 1814 mark was shown in 1815, and during the next decade the total amount started to climb sharply, going over 35,000 tons in 1818 and 1819.[7] The low figures from 1812 to 1815 indicate clearly that American ports were caught short of their expected coal supply.

Unfortunately, federal statistics do not give the amount entering each American port. According to the Mutual Assistance Coal Company, Philadelphia was accustomed to receiving about 3,000 tons of

Virginia coal annually.[8] By April 1813 the severe shortage of coal in Philadelphia caused the price to triple. The city's craftsmen faced a fuel crisis which resounded throughout the economy. These artisans had to pay high prices for the coal they acquired and passed the extra cost on to the consumers. Psychologically the city was stunned as it turned to Pennsylvania anthracite for relief.

In addition to the letter he received from James Ronaldson, on 5 April 1813 Jacob Cist heard similar news about the fuel crisis from Daniel Freytag, a Philadelphia potter who used coal to heat his kilns. He wrote to Cist about the opportunity of introducing anthracite into the Quaker City. Believing that the British blockade would last at least a year, he said that Virginia coal sold at $1.00–1.25 per bushel. If Cist agreed to send a wagon load of Wyoming Valley coal to Philadelphia, the potter would assist in finding a market for it. Before a market could be established, he suggested distributing the anthracite to different shops. Once a market had been created, the potter felt there would be no difficulty in selling hard coal.

Jacob Cist and other merchants believed that Wyoming Valley coal would answer the needs of Philadelphia. Several four-horse wagons loaded with 1.5 tons of anthracite left the Wyoming Valley for southeastern Pennsylvania. In 1813, Cist's brother-in-law George M. Hollenback sent two loads to Philadelphia, and the owner of a Nanticoke mine shipped one load to a blacksmith in Germantown.

James Ronaldson was the first Philadelphia craftsman and merchant with whom Jacob Cist made arrangements to market Wyoming Valley coal. In June 1813 Cist shipped a wagon load of coal to Ronaldson. Fearing that high prices might tempt the wagoner to sell the coal en route, and to guard against mischief, Ronaldson advised sending another load immediately and paying the wagon master only $4 or $5 on account. He felt that a regular receipt in the form of a bill of lading would provide for safe delivery.[9]

On 6 July 1813 Ronaldson received another 2.5 tons of anthracite from Cist. Some men at the typefoundry succeeded in using hard coal, but it was still too soon to give any opinion on how well Cist's coal would answer purposes. The typefounder requested another load. Three days later a wagon carrying 1.5 tons left Wilkes-Barre for the foundry. The driver received $20 a ton for freight upon delivery. The 120-mile trip took six days. In August Cist shipped another ton to the Cedar Street foundry. Ronaldson liked the looks of the large chunks that came in the August load.[10] In Philadelphia Jacob Cist had won a steadfast customer for his Wilkes-Barre coal. It melted lead quickly for casting type. At the beginning of September James Ronaldson asked Cist for eight or ten more wagon loads of anthracite.

He wanted to stock up before cold weather set in and freight costs increased even higher.

Ronaldson also felt that the time had arrived for introducing Wilkes-Barre coal into the Philadelphia market, and winter would bring more buyers. He agreed to sell Wilkes-Barre coal on commission if Cist financed the business. The typefounder warned Jacob that the compensation would bear no proportion to his trouble, but everything could be arranged for mutual satisfaction. During the fall of 1813 Cist recorded sending 12 wagons carrying a total of 20 tons to Philadelphia. Ronaldson said that Wilkes-Barre coal sold at $1 a bushel. He was happy that others tried this coal because as more Philadelphians bought it, greater attention would be given to the anthracite trade and eventually measures would be taken to establish a steady, dependable supply. The typefoundry could be certain of receiving anthracite and served as an outlet for Cist's Wilkes-Barre coal.

The logical route for shipping Wyoming Valley coal to market was down the Susquehanna River. Nor did Jacob Cist overlook Baltimore as a possible market for his anthracite. Due to the British blockade, the fuel crisis hit Baltimore the same month that it affected Philadelphia. On 12 April 1813 William L. Brockelman, a Baltimore merchant, informed Cist that the blockade caused a rise in the price of Virginia coal to $.50 a bushel, and he encouraged Cist to send anthracite to his city.

Brockelman's proposal appeared most attractive. As long as the British navy didn't lift the blockade so that coal prices dropped to the normal $.30 a bushel level or cut trade between the Susquehanna and Baltimore, conditions were good for going into the coal business. If Virginia coal remained at $.50 a bushel, it would pay to send coal from Wilkes-Barre. Cist had told him previously that Wyoming coal sold for $.08 a bushel or $2.25 a ton on the riverbank at Wilkes-Barre. Brockelman wanted to know how much it would cost to deliver anthracite at Havre de Grace. He hoped to sell it for the prewar price and planned for trials at a glasshouse, foundries, and blacksmith shops. He asked Cist to send an ark load of 35 or 65 tons of the best Wilkes-Barre anthracite down the Susquehanna and let him know how long it would take the coal to get to Havre de Grace so that a ship could be ready to pick it up. The owner of a Baltimore glass factory agreed to join him in the venture. Should anthracite gain acceptance in Baltimore, a large amount would be needed from Cist's mine. Success could mean the establishment of a coal yard in which the Baltimore merchant offered Cist a partnership.[11]

The coal business was not new to Wyoming Valley residents. Like Abijah and John Smith, Cist could send anthracite down the Susquehanna River. John Smith used arks 90 feet long, 16 feet wide, and 4 feet deep to carry 40–60 tons of hard coal to market. Susquehanna raftsmen possessed excellent reputations for taking arks down river with little mishap on spring and fall freshets.[12] From June 1812 to January 1813 two New York City commission merchants sold over 200 tons of Smith's coal in their city. While the Smith brothers led the pack, they were not the only businessmen in the valley who engaged in the Susquehanna coal trade. Jacob Cist welcomed the opportunity to establish a market for his anthracite in Baltimore.

Cist sent Brockelman an extensive statement concerning the coal business. Annexed to a letter of 2 May 1813 were strict instructions for burning hard coal. In the next letter Cist enclosed drawings of anthracite grates. By the beginning of May over 50 tons of Wilkes-Barre coal awaited the Baltimore merchant at Havre de Grace. Cist also informed him that anthracite could be purchased from Abijah Smith. During May and June the British squadron kept ships from bringing Cist's anthracite to Baltimore. Several tons of coal were carried by wagon to the Baltimore glass furnace. Not until July did the navy move down the Chesapeake Bay and allow Cist's shipment to reach its destination.[13]

The desire to burn hard coal in the Baltimore glass furnace failed. The fire had to produce large flames which came out of holes in sides of the furnace where men blew glass. On their first trial neither Jacob Cist's nor Abijah Smith's coal made sufficient flames to manufacture glass. The Baltimore merchant visited a Pittsburgh glassmaker who was familiar with Wyoming coal. The glassman said that if anthracite came from pits at least 100 feet deep, it would contain enough gas to give the proper flame. However, no mine in the Wyoming Valley went that deep. The merchant remained willing to try more anthracite in the glass furnace. On 28 March 1814 he informed Cist he would take 2,000 bushels of the best Wilkes-Barre anthracite. His partner would pay the bill.

The merchant's attempt to market anthracite in Baltimore met with better success than his experiments in the glass furnace. In May 1813 he estimated that Baltimore consumed from 7,000 to 10,000 tons of coal annually. He hoped to establish a coal yard to handle business. Blacksmiths in particular should be interested in anthracite. On 1 November he told Cist that Susquehanna coal sold for $.75 a bushel and yielded a handsome profit. The price of coal at Baltimore included $135 for 60 tons at Wilkes-Barre, $50 for the ark, $60 for the Susquehanna trip, and $100 for freight from Havre de Grace. The cost came to more than $345 since other expenses such as storage must be added. At $.75 per bushel the load sold for $1,320 and provided a profit of nearly $1,000. The Baltimore merchant hoped the high quality of anthracite would become better known so that sales would increase. Public houses burned some anthracite, and the merchant thought it would gain popularity the ensuing winter. On 28 March 1814 he reported to Cist that he inquired periodically about progress made by Susquehanna coal in Baltimore. He found little hope for its successful adoption. Although several blacksmiths used anthracite, they preferred Virginia soft coal. He attributed their dislike mostly to prejudice. Recently, several cargoes of Virginia coal had escaped British vigilance and reduced anthracite prices from $1.00 to $.60 per bushel. The supply from Virginia remained uncertain and kept prices fairly high. While Virginia coal was the favorite fuel among Baltimore residents and craftsmen, Pennsylvania anthracite, thanks to the efforts of Cist and others, had found a market in that city. Cist's dream of establishing an anthracite trade was slowly coming true.

Wilkes-Barre also served as the center for initiating the Lehigh coal trade during the war. The Baltimore merchant with whom Cist dealt originally had considered Lehigh coal for his supply. On 21 April 1812 he informed Cist that he had visited the coal mines on the Lehigh River north of Bethlehem and discovered more difficulties in the probable success of getting coal to market than he imagined. Unless the Lehigh Coal Mine Company embarked on an expensive undertaking, he stated, nothing could be done.[14] At the end of the year a Delaware County iron manufacturer came to the same conclusion. Needing coal to operate his rolling mill, he visited Jacob Weiss to ascertain the prospect of getting a regular supply of anthracite from the Summit Hill mine. He obtained information from Weiss, explored the margin of the Lehigh River for some distance, and went up to see the mine worked by the coal company. After inspecting the coal bed and the approach to it, the iron manufacturer examined the surrounding terrain to see if he could construct a better road to the river at moderate expense. The rough aspect of the country gave him little encouragement, and he found mines on the Schuylkill headwaters better located to supply him with anthracite.[15]

Jacob Cist was in a better position to initiate the Lehigh coal trade even though his uncle had kept the mine producing on a limited basis since the 1790s. The Cist-Hollenback firm operated on a far larger basis than Jacob Weiss's enterprise. From Wilkes-Barre Jacob Cist could supply the men and material to keep the Lehigh coal mine in continuous production. Since Cist had to spend most of his time in Wilkes-Barre taking care of his father-in-law's business, he needed the able assistance of someone to reside near the Lehigh coal mine and oversee the operation. He counted upon Isaac A. Chapman, a close friend who lived in Susquehanna County, to become his most trusted partner in this venture.

One month before the War of 1812 began, Chapman wrote to Cist about plans for rafting coal down the Lehigh River. Chapman said he could not begin operations for two months. On 12 July 1812 Cist and Chapman became partners in the Lehigh coal trade. Cist took charge of mining and carting to the riverbank as much coal as he considered proper. Chapman assumed responsibility for acquiring timber, employing men to construct arks, and conveying coal to market in Philadelphia and elsewhere. Profits from all sales would be proportionally divided between the two partners.[16]

Before commencing business Cist and Chapman waited until the fuel crisis occurred in March 1813. On 14 March Cist wrote Chapman at Lausanne, their center of operations on the upper Lehigh for conducting the coal trade, that he was waiting in Wilkes-Barre for two merchants to fill a contract. He told Chapman to get planks for four boats and that 40 tons of coal would soon be delivered on the riverbank. Three men and a boy dug coal at the mine. While a blacksmith stayed at the mining camp to mend tools, three teams carted coal to the river. Chapman would find towline, ropes, and spikes for rafts in the garret of a white house at Lausanne. As soon as water permitted, he would take two men and a pilot and run a couple of boats to market.[17] Apparently, they sold the coal. James Ronaldson may have purchased some of it because during the summer he ordered anthracite from Cist in Wilkes-Barre. By the winter of 1813–1814, the increased demand for hard coal in Philadelphia encouraged Cist to enlarge his Lehigh operations.

In 1813 Cist calculated costs for conveying Lehigh coal to the Quaker City. He estimated expenses by using data relevant to the Wyoming Valley and figured 30 bushels equaled one ton. Coal at the bed was valued at $.90 per ton, and the cost for digging coal amounted to $.60 per ton. To open a new bed in Luzerne County, miners received $1 per ton for the first 100 tons and $.50–.60 for every ton thereafter. At this rate Jacob Cist said colliers made very good wages. He figured it would cost $1.80 a ton to transport anthracite four miles from a mine near Nesquehoning to Lausanne. A two-horse team averaging one ton per load could make two trips a day and earn the drivers excellent pay. He added $.45 per ton for handling and putting coal in arks. An ark 70 feet long which carried 800 bushels, or 26 tons, cost $90.00. Cist put down $45.00 for an ark because he could resell it in Philadelphia at half the original price. To run an ark required four men and took one week. At $1.25 a day per person labor for arking came to $35.00. River transportation amounted to $3.00 per ton. Insurance on the ark and coal ran $.75 per ton. Expenses per ton totaled $7.50.[18] If Cist's calculations proved correct, Lehigh anthracite could compete with prewar prices on other coal. At the prevailing high prices, possible profits served as a strong inducement to attract large sums necessary to undertake the enterprise.

Additional capital for the Lehigh coal trade came from other Wilkes-Barre businessmen. On 15 December 1813 Jacob Cist, Charles Miner, and John Robinson, a Wilkes-Barre merchant, received a special ten-year privilege from the Lehigh Coal Mine Company. The Wilkes-Barre entrepreneurs obtained rights to dig and take coal from the mines in common with the company and their assignees. The company retained the power to grant similar privileges, such as the one held by Cist and Chapman, with other individuals at pits opened during previous decades. Cist and his partners received exclusive rights to the mines they opened and could explore for sites anywhere on the 10,000 acres. Roads and landing places would be used jointly by everyone. Cist, Miner, and Robinson could cut timber from company land for building arks, making oars, and erecting any buildings that were necessary to carry on business. Upon termination of the agreement, buildings would revert to the company. In return for these privileges Cist and his friends agreed to deliver to Philadelphia at least 5,000 bushels annually during the first two years and 10,000 bushels thereafter. They paid a token fee of one barleycorn annually for the coal.[19] Initially the Lehigh Coal Mine Company received little from the agreement. Since Cist owned four shares of stock, one company owner benefited. In a way the company repaid the Cist family for the work of Jacob's father. If the enterprise succeeded, the company ultimately gained from increased value of Lehigh anthracite and its coal land.

The same day the Wilkes-Barre partners received special privileges, Cist let the Lehigh company managers know about arrangements that were decided upon for doing business. Cist, Miner, and Robinson agreed to determine whether they could profitably take anthracite to market. They believed a full and fair experiment would take until the following spring. They would share costs jointly, advancing equal amounts from time to time as deemed necessary for doing business. Each partner received an allowance of $.50 per day for the time he spent on the coal business. Profit or loss would also be equally shared. They would keep a journal, recording all money spent and transactions made. In case the partners found business profitable, they agreed to make arrangements for prosecuting business on an extensive scale.[20]

John Robinson took charge of operations on the upper Lehigh. On 4 January 1814 he received $250 cash and headed for Lausanne. During the next three months he faced continuous trouble. Deep snow and frost forced the partners to abandon plans for opening a new mine at Room Run four miles above Lausanne; consequently, they depended upon the Summit Hill mine for coal. Robinson wrote to Cist in February that so far the biggest difficulty was transportation. Although the snow had melted, an upper road via Mauch Chunk Creek remained so wet and slippery that no team would use it until dry summer weather arrived. Only Weiss's road remained usable. Robinson had examined the ground between the Summit Hill mine and another coal pit for a possible road. He planned to hire several hunters to help find a path and hoped to discover a route from Summit Hill via Nesquehoning Creek to Lausanne which would shorten the hauling distance by one or two miles. Poor roads continued to plague Robinson. In March he reported that roads were so bad it was almost impossible to haul coal.[21]

On 21 February 1814 John Robinson informed Cist that the time had come for wagons to start carrying coal. Finding it difficult to get wagons without paying extravagant wages, he wished he had some Wyoming Valley teams. He had spent the previous day riding through Mahoning Valley trying to find teams but did not know whether anyone would show up to wagon coal. By mid-March two teams kept coal moving from the mine to the landing. Robinson accompanied the wagons for two days and determined that 35 bushels made a full load. To prevent the wagoners from quitting, Robinson paid them more than $4.00 a day. Costs for carrying coal from the mine to the landing averaged $3.00 instead of Cist's previously predicted price of $1.80 a ton.

Another problem was constructing arks. In February Robinson wrote Cist that the man who supervised the ark yard remained hopeful about taking coal down the Lehigh River. Robinson, who spent most of his time getting timber and preparing for the trip downsteam, felt optimistic initially about this phase of the business but within a month had changed his attitude. On 12 March he paid for the ark timber and set out for Wilkes-Barre to find his ark yard foreman. Upon arriving in the Wyoming Valley, he discovered the foreman had abandoned their Lehigh enterprise. Robinson said the foreman's conduct had helped ruin the experiment, but somehow the Wilkes-Barre partners acquired five arks at $100 apiece. Each ark could convey 20 tons of anthracite.[22] Once again actual expenses went above Cist's original calculation.

Ultimately John Robinson's greatest problem concerned mining coal. On 21 February 1814 Robinson wrote Cist that two men had contracted to raise 100 tons of coal for $100. Robinson felt relieved from worrying about mining coal, but his optimism regarding this part of the business soon disappeared. On 4 March a relative of Charles Miner made an agreement with Jacob Cist to dig at the mine 200 tons of anthracite for $200 and agreed to hire a helper.[23] Cist also promised to pay him liberally for as much coal as he could mine above the quantity stipulated in the contract. To accommodate this laborer and another party, Robinson bought out the previous mining contract. Plans fell through when the second party decided to engage in logging instead of mining and when one of Robinson's former colliers went home. Fortunately, Robinson found another man to dig coal. His biggest disappointment, however, came from Charles Miner's relative who was needed to keep one of the two wagons supplied with coal. The day after arriving at the coal bed, he attempted to blast in the rain, failed, got angry, and left for home without telling Robinson. The water in the coal bed was very troublesome. To drain the mine required sinking a ditch eight or ten feet deep. Additional coal diggers could not be found. Robinson especially blamed Miner's relative for the project's failure but praised one collier for sticking to his contract.[24] One hundred tons waited on the riverbank for shipment downstream.

While John Robinson supervised operations on the upper Lehigh, Jacob Cist took charge of supplies and selling the coal. Provisions came from Matthias Hollenback's Wilkes-Barre store. Cist went to Lausanne several times to inspect the work and reported progress to G. Haga, a friend who was president of the Lehigh Coal Mine Company. Cist asked him the following questions about marketing coal in Philadelphia. Were Steelwagon and Knight, commission merchants, able to handle anthracite? Were they men of standing and capital who would order a large quantity? Would they pay teamsters for overland carriage upon delivery in Philadelphia? How much would they pay for a specific quantity of anthracite delivered in the city within a given time? Cist told his friend to expect a large quantity of Lehigh coal in Philadelphia within a short time.[25]

By 19 March Cist had sold the coal and told John Robinson to be ready for the trip down the Lehigh River. Six days later the Wilkes-Barre partners settled accounts on their coal business. Operating costs came to $10.50 a ton, $3.00 above Cist's original figure. Expenses amounted to $1.00 a ton for mining, $3.00 a ton for hauling, $5.50 a ton for arking, and $1.00 a ton for provisions. One hundred tons sold for $2,700.00, yielding a profit of $1,650.00; thus each partner received a profit of $550.00. Cist recorded group reaction as one of "Comic Tragedy performed with great applause on the Lausanne Theatre" and listed the cast of characters.[26]

As previously agreed, the Wilkes-Barre partners formed a new arrangement based on experience. They lost the support of John Robinson, who sold his share to a New Haven, Connecticut, merchant. In June 1814 the New Haven merchant and Charles Miner became partners with Miner assuming managerial responsibilities while Jacob Cist handled his part of the business, that is, mining, hauling, and selling coal.[27] Since Robinson's position as superintendent of operations remained vacant, Cist turned to Isaac Chapman for some help. The partners relied on subletting contracts to get work done.

On 1 April 1814 Charles Miner's relative took another advance on his contract to mine coal and went back to work at Summit Hill. For the next seven months he dug coal. Assuming fulfillment of the contract, he put 200 tons in a convenient place near the pit by October.[28] Isaac Chapman, who joined him at Summit Hill in April, had dug 500 tons by 1 September. The task of taking anthracite from the mine had been solved as coal continued to come from the quarry. Chapman also opened the mine at Room Run near Nesquehoning and kept it producing. Although they do not give a total figure, accounts show steady coal production throughout the War of 1814. One other mine added to total production in the Lehigh region. In 1813 Nathan Beach, a good friend of Jacob Cist who lived on the Susquehanna near Berwick, opened a mine at Beaver Meadows and sent coal to the landing at Lausanne.[29]

Coal piled up near the mine at Summit Hill. During the spring of 1814 a few wagons took coal from the heaps, but most of it remained there until July. On 15 July Jacob Cist advertised in Germantown and Philadelphia for wagons to haul between 120 and 140 tons of coal over a rough eight-mile downhill road from Summit Hill to the mouth of Mauch Chunk Creek. Cist offered to pay $3 per ton if wagon masters finished the job by 1 September.[30] Evidence indicates that no one accepted Cist's offer. Pressure to get more coal to the riverbank forced Cist to raise the price for hauling coal. On 8 August Cist made a contract which paid $4 per ton to carry from 15 to 20 tons to the mouth of Mauch Chunk Creek. Wages for hauling coal remained at the $4 a ton mark for the war's duration.

At the higher price Cist received sufficient help to haul about 700 tons to the riverside during the fall. Isaac Chapman took 10 tons to Mauch Chunk, and Charles Miner's team carted a similar amount. John Robinson's team earned $48 for carrying 12 tons to the river.[31] The amount of coal that was mined stayed ahead of the volume conveyed to the Lehigh. In December Cist made contracts to carry 500 tons from the "great coal mine" to Mauch Chunk. The contractor agreed to put six to ten wagons on the coal road. He could also dig coal at $6 per ton to get a sufficient quantity for completing the contract.[32] Cist aimed at getting at least 1,000 tons to market during that winter.

Jacob Cist brought teams from the Wyoming Valley to keep anthracite moving down the Summit Hill road. On 18 January 1815, a Luzerne County wagon master agreed to haul 60 tons from the great mine to Mauch Chunk before 1 April. If he fulfilled the contract, Cist would pay $4.50 a ton haulage. Any amount short of 60 tons required him to pay only $4.00 per ton. During the next month Cist made two more contracts of a similar nature, one with a wagoner from Luzerne County and another with a Northampton County resident. Both agreed to haul 60 tons by 1 April. As soon as the snow fell, Isaac Chapman started the teams previously engaged to cart coal, went to Lausanne, and acquired every sled he could possibly get. During February Chapman supervised the transportation of approximately 200 to 300 tons of coal by sled down the snow-covered road.[33]

By the spring of 1815 huge heaps of coal lay along the bank of the Lehigh River from Lausanne south to Lehighton. Lausanne proved to be a poor port from which to ship coal. A sharp curve in the river between Lausanne and the mouth of Mauch Chunk Creek two miles to the south was dangerous for rafting. Anthracite sent from the Beaver Meadows and Room Run mines traveled shorter distances overland to the Lausanne landing and was sent downstream from this point. Most of the anthracite mined during the war came from the quarry at Summit Hill and went to the river via a road along Mauch Chunk Creek. Accounts and contracts called the landing at the mouth of that creek Mauch Chunk, or Coalville.[34] Mauch Chunk became the logical place to ship coal down the Lehigh. Broad flat plains along the west bank of the river two miles south of Mauch Chunk provided additional space to store coal. And even farther to the south Lehighton served as a site for doing business. Coal could be brought down a road running from the Summit Hill mine through Mahoning Valley to Lehighton. In 1814 Lehighton was laid out as a town in order to carry on the coal trade. Isaac Chapman copied plans of the town and considered it as a spot for unloading coal from wagons.[35]

Chapman and Cist assumed responsibility for marketing anthracite in the Lehigh Valley. On 12 April 1814 a Bethlehem storekeeper wrote to Jacob Cist about the coal business they had discussed previously. Another Bethlehem merchant was willing to join him in attempting to introduce Lehigh coal. He told Cist to send three tons on a raft. This type of transportation would reduce costs to a point where the price would be equal or lower than wood. A competitively low price was one of the major prerequisites for introducing anthracite in the Lehigh Valley. The Bethlehem merchant offered to pay a reasonable price for putting coal on the raft, conveying it downstream, and unloading it on the shore. He sent a similar letter to Cist's cousin, Thomas Weiss, at Fort Allen.[36]

Three weeks later the Bethlehem storekeeper increased his order for anthracite to a boat load of 10 or 12 tons. He said transportation by boat to the Bethlehem bridge would cost $3 a ton. He wanted a full quantity and hoped Cist would not disappoint him. A German boatman had promised to fetch the coal on the next freshet, and Cist should be ready to send coal on a moment's notice as soon as it rained. Eventually anthracite went to Bethlehem and gained acceptance. In August Cist contracted to haul as much coal as he could from Summit Hill to Bethlehem at $10 per ton.[37] Increased demand for anthracite in the Moravian town made overland shipment worthwhile.

Christian Hornick, an Easton merchant, informed Jacob Cist that it did not pay to transport coal by boat to his town because expenses for a boat amounted to more than the cost of sending coal by wagon. In June 1814 Cist sent two wagon loads of coal to another Easton merchant and six months later a boat carrying over eight tons of coal headed down the Lehigh for him.[38] Most of the time Cist shipped coal overland to villages in the Lehigh Valley. During the winter sleds supplemented wagons.

Cist's customers fell into two groups. Iron workers became steady consumers. Samuel Felter, a Bethlehem blacksmith, purchased coal three different times, once in August 1814 from Charles Miner, a second wagon load in December, and five more sled loads the following February. Cist sent coal to two gun manufacturers who had burned anthracite since the 1790s. On 17 August a Nazareth gunsmith requested a load of 35 bushels that were promptly delivered to his shop.[39] Six months later he received two more loads. This gunsmith also used Wilkes-Barre coal for fuel. Although he preferred Lehigh anthracite, he could get coal from the Wyoming Valley at a cheaper price.

Jacob Cist traded anthracite locally for supplies. One Easton merchant gave him a wheelbarrow, a scale beam, eight chains, and 132 pounds of spikes as partial payment for 1.5 tons of coal. From 28 May 1814 to 7 February 1815, a Bethlehem merchant received nine loads of coal and paid part of his bill with 32 bushels of rye, 12 bushels of oats, one horse collar, and four straps.

An Allentown merchant used a stove invented by Jacob Cist to heat his store. On 1 January 1815 Francis Weiss wrote Cist that the stove consumed about four bushels of anthracite a day. Such a stove burned 15–20 tons a year. Since few people in the Lehigh Valley could afford the price of an anthracite fire, Cist's coal went primarily for other purposes. During 1815 Cist sent several wagon loads to villages in the Lehigh Valley. In this manner he distributed approximately 75 tons of coal and expanded the market started by the Weiss family.

Some Lehigh anthracite went as far as Philadelphia by wagon. On 15 July 1814 Jacob Cist posted handbills in Germantown and Philadelphia calling for three teams to cart coal from the Lehigh River back to the city. Cist said he would pay $14 per ton freight. The round trip would take about four weeks. Cist suggested that teamsters take full loads of feed on their journey north, since it was needed for feeding animals at the mine.[40] Whether anyone accepted Cist's offer remains unknown. The Lehigh Coal Mine Company account book shows no entry for overland freight to Philadelphia. Since wagons carried coal from the head of the Schuylkill River and Wilkes-Barre, it seems safe to assume that a few wagons also traveled to the Lehigh coal mines.

Joseph Smith, a blacksmith living at Tinicum, 20 miles south of Easton on the Delaware River, received four loads of coal via land transportation. In February 1815 two sleds carried a total of 1.5 tons to Tinicum. The blacksmith's wagons picked up two more loads at the Lehigh coal yard in April. He wanted anthracite so much he would rather pay high overland freight rates than use charcoal.[41] Most of the time he acquired coal from arks descending the Delaware.

The problem of constructing arks and rafting anthracite provided the greatest trouble for Jacob Cist and his partners. The most expensive part of the operation involved river transportation, and money was extremely scarce during the war. To acquire sufficient funds, Cist made several contracts with other businessmen to build arks and transport coal down the Lehigh and Delaware rivers to Philadelphia. In every contract Cist agreed to mine and haul coal to the Lehigh River. His partners assumed responsibility for getting the coal to market.

As partners in the enterprise, Charles Miner and other merchants undertook the task of moving anthracite from the Lehigh landing to Philadelphia. In June 1814 Miner took ark builders, workmen, teams, and provisions from Wilkes-Barre and went to Lausanne.[42] A sawmill at the mouth of Nesquehoning Creek was leased from Jacob Weiss. Carpenters constructed arks 65 feet long by 14 or 16 feet wide which drew 20 inches of water while carrying 20–30 tons of coal. Miner estimated the construction of an ark cost $130, nearly 50 percent higher than Cist's original calculation.

On 12 July 1814 Chapman agreed to build arks and transport coal for Cist.[43] Jacob needed arks before the boating season arrived, and in July he had tried to get an Easton carpenter to build boats for him. Although the Eastonian was too busy to make an agreement, he promised to construct several arks the following winter. Cist made contracts with several men to construct arks and counted on having at least 12 for taking coal to market.[44]

During the first week of August 1814, rain raised water in the Lehigh River to arking depth. On 9 August Jacob Cist dispatched Charles Miner's first ark down the Lehigh from Lausanne. A pilot and five assistants steered the large craft down the wild rocky stream.

The sharp turn between Lausanne and Mauch Chunk caused considerable trouble. Within a quarter mile of the landing the ark struck a ledge and broke a hole in its bow. The boatmen stopped the leak by plugging the hole with their clothes. By dusk the craft had traveled 50 miles to Easton. However, the trip down the Delaware went much slower. A pilot and two assistants required four days to go from Easton to Philadelphia.[45]

Recalling the venture 30 years later, Charles Miner said the task of arking coal was dangerous. He sent down a number of coal arks and claimed that three out of four either stove or sank in the river. Attributing the difficulty to oversized boats, he felt that crafts which carried 10 tons instead of 25 could be taken down the Lehigh with much less loss. Upon arriving at Easton, several small arks could be tied together for the trip down the Delaware. Fortunately, during periods of low water Miner was able to recover coal which had sunk in the Lehigh. As water receded, he moved the coal to the riverbank and sold it to local blacksmiths.[46]

Evidence indicated that Cist's partners never lost as many arks as Miner claimed. Isaac Chapman left the Lehigh landing with an ark load of coal the day before Miner's anthracite went downstream and arrived safely in the city. In their accounts of the coal trade, neither Chapman nor Cist mentioned losing coal arks.[47] Their correspondence contains only one reference to a wrecked ark. Most of their coal arrived safely in Philadelphia.

Miner's reminiscence distorted another major fact about the early coal trade. He said the cost for his first load of 24 tons was $330.77 but neglected to mention it sold for $2,400.[48] A profit of $2,000.00 more than paid for the expense of the enterprise. Chapman's ark of anthracite earned equally high profits.

Upon returning to the Wyoming Valley for more supplies, Isaac Chapman stopped at the store of a Wilkes-Barre merchant who quizzed him about the Lehigh River and coal business. They went into a back room where the merchant requested Chapman's opinion about the chances a young Wilkes-Barre businessman would have of succeeding in the trade. Chapman liked the idea. The young businessman was someone he knew and upon whom he could depend. The merchant consulted the young man's mother, who consented to her son's engaging in the business.[49]

Chapman gave the merchant the usual terms of a contract. Cist would agree to deliver coal at the river while the merchant and his young partner would have to acquire timber, build arks, and take the anthracite to market. Profits would be shared with 50 percent for Cist and 50 percent for the partners. Chapman told the merchant he would deliver 1,000 tons annually on the riverbank until the lease from the Lehigh Coal Mine Company terminated. The Wilkes-Barre merchant expressed satisfaction with the terms.

Before signing a contract, Chapman wanted the two partners to go to the Lehigh and examine everything for themselves.[50] They went to the Lehigh and examined the mine, road, river, and timber. The appearance of the anthracite and extent of the mine satisfied them. Furthermore, the road was in better shape than expected. Desiring to take 1,000 tons to market annually, the merchant asked for the 500 tons at the mine and wanted more coal dug immediately. He feared Chapman could not get 1,000 tons to the river annually. The partners rode along the Lehigh to Weiss's home, examined the river, and felt confident the young businessman could run 75-foot arks with ease by Mauch Chunk bend. After much discussion the merchant said he had made up his mind to engage in the Lehigh coal trade and expressed confidence in the prospect of making money.

The Wilkes-Barre merchant wanted Chapman to let his partner have the arks, timber, and coal so that he could immediately step into the business. Chapman agreed as long as they paid fair compensation for his expenses and time. On 6 September 1814 they signed a contract with Jacob Cist to transport Lehigh anthracite to market. Cist told them he would deliver annually from 300 to 1,000 tons at the Lehigh landing.[51]

The youth did not start working as quickly as his partner anticipated. The month of September passed before work commenced. In October he went to Lausanne, contracted for plank and timber for 20 additional arks, cut logs and oar sticks, and cleared land for another ark yard. He employed a carpenter and three helpers to build three or four arks, which were as many as he thought prudent to construct at that time, owing to the lateness of the season. As soon as water in the Lehigh became suitable for navigation, he ran the arks and coal purchased from Chapman to Philadelphia. He returned to the upper Lehigh, built two more flats, and took them to market. After finishing this business, he decided to complete several ark bottoms which would be put in the water in the spring.

The delay of the two Wilkes-Barre partners during September 1814 caused Cist to fear forfeiture of his lease with the Lehigh Coal Mine Company. By September Cist and his partners had taken about 150 tons of Lehigh coal to Philadelphia. They remained at least one ark load shy of the 5,000 bushels stipulated in the lease. If for some reason anthracite did not reach Philadelphia in the fall, they were vulnerable to losing their lease.

On 18 October 1814 Jacob Cist wrote to a stockholder in the Lehigh Coal Mine Company about the possibility of waiving the

5,000-bushel provision. Without consulting his colleagues, the stockholder replied confidently that the request would meet with company approval. At the end of October the company resolved to disregard the stipulation.[52]

Meanwhile Isaac Chapman considered competing with the two Wilkes-Barre partners in the Lehigh coal trade. He proposed forming a partnership with two of his merchant friends at Binghamton, New York. With their capital Chapman could fulfill a contract made with Cist. On 18 October 1814 Joshua Whitney, one of the Binghamton merchants, wrote Cist about their plans. He thought the partnership would be worthwhile if the quantity and quality of Lehigh coal proved sufficient and navigation safe. His partner planned to inspect the operation, ascertain the prospect of success, and consult Cist's partners on the subject of uniting all interests into an association. The merchant felt the plan would prove highly advantageous by eliminating rivalry and giving energy to the whole system. He wanted Cist's opinion about the plan.

Three weeks later Whitney received Cist's reply to his letter and his partner's report. The report was so encouraging that the merchant decided to support the project. He told Cist he had sold out his share of goods and was making speedy arrangements for settling accounts. He expected to raise a sizable fund for investing in the coal business. In November he sent his youngest brother to learn how to navigate the Lehigh River. His brother was reported to be a skilled Susquehanna River pilot. After a trip down the Lehigh River, the merchant said his brother would become a valuable pilot for the trade. The Binghamton merchant did not feel he could get to the upper Lehigh until the following February. Thereafter he would spend most of his time in the coal business.[53]

On 27 November 1814 Isaac Chapman asked Cist about his plans for the ensuing winter. Had Cist made arrangements with the Binghamton businessmen? Chapman anticipated serving as their agent on the upper Lehigh. Much to his surprise, nine days later Chapman learned that one of the merchants had abandoned all intentions of engaging in the coal trade. Chapman gave Cist two reasons for the merchant's decision: poor health and a misunderstanding about Cist's shares in the Lehigh Coal Mine Company. One partner had led the other to believe that Cist would purchase shares of the Lehigh Coal Mine Company and resell them to the proposed company. Upon learning that Cist had bought several shares with which he did not intend to part, the merchant decided to withdraw his support. Chapman said he never heard of such an arrangement and felt that someone had misinformed the Binghamton merchant.[54]

The merchant's decision disappointed Chapman. Without his

money Chapman's desire to pursue the coal business on a large scale was frustrated. Although the other merchant still wanted to engage in the trade, he possessed insufficient funds to enlarge operations. Chapman told Cist that he would approve any plan by which funds could be procured at a reasonable rate and in such a manner as to make acceptance of them safe. Unless a sufficient sum could be obtained upon terms similar to those made by the Binghamton merchant, Chapman did not want to borrow money.

Banks refused to lend money to the two anthracite entrepreneurs. Henry D. Mandeville, an official of the Northampton Bank at Allentown, wrote Cist the sad news that the directors turned down his request for financial help even though the more enlightened managers wanted to encourage the coal trade. Greater demand for money in the lower Lehigh Valley prevented compliance with their request.[55]

In Philadelphia Steelwagon and Knight had expanded their coal trade to handle anthracite for Jacob Cist and his partners. Customers could pick it up at their store at 219 North Second Street or have it delivered. Coal generally sold for cash, although a few customers purchased anthracite on credit with four months to pay.

Demand for Lehigh anthracite in the Quaker City fluctuated with supply, price, and the habits of consumers. Steelwagon and Knight sold the coal that arrived in August 1814 fairly quickly. A manufacturing company purchased a large quantity at $.75 per bushel on four months' credit. The 500 tons of anthracite which the Wilkes-Barre partners sent down during the fall glutted the market. Soft coal consumers needed to be taught the methods of using anthracite before they would purchase Lehigh coal. (Cist's campaign to sell Lehigh anthracite will be discussed in chapter 3.) An equally important reason for slow sales during the fall of 1814 was the high price. In refusing to purchase Lehigh coal, consumers tried to force the price below $1 per bushel. Cist wrote Steelwagon and Knight on 3 December to keep the price above $1 a bushel. They complied with his request even though anthracite piled up in their yards.[56] The arrival of winter brought a change in consumer attitudes as blacksmiths paid the high price for their fuel supply. At the end of the year about 70 tons of Lehigh coal remained unsold in the city.

Coal sales during the first four months of 1815 remained steady. On 15 February Steelwagon and Knight reported that they had disposed of 18 tons, about half of which was for cash. They also informed Cist that peace at last had been concluded. Hostilities would cease as soon as President Madison signed the treaty. The blacksmiths expected to get coal from Virginia shortly at the prewar price. The coal merchants feared Lehigh coal would not sell so readily.[57]

News of peace destroyed the ambition of the two Wilkes-Barre partners to send coal to market. While 240 tons waited at the Lehigh landing for them to send to Philadelphia, no arks went down the Lehigh. One partner told Chapman he did not feel like taking any more funds from his other business. He offered to run the anthracite on Chapman's account.[58] In April 1815 Cist informed Chapman that he wanted to send 200 tons downstream. But they did not use the spring freshets to get anthracite to market. Cist wrote to Chapman on 2 May that he had just received the account of coal sales from Steelwagon and Knight. The price since peace had been $.75 a bushel, which was higher than he expected. Steelwagon and Knight had sold all the Lehigh coal in their possession. In addition to James Ronaldson, Cist and his partners succeeded in finding another mercantile firm to market Lehigh coal in the Quaker City during the war. And they were willing to continue doing business with Cist after the war.

Philadelphia was not the only American city in which Jacob Cist attempted to introduce Lehigh anthracite. On 14 November 1814 Charles Miner wrote Cist from New York City about his efforts to sell anthracite. New Yorkers who had seen Miner's specimens of Lehigh coal were pleased with it and wanted one or two ark loads. Miner expressed disappointment that the ark load which he ordered had not arrived and gave Cist instructions about sending it to New York. Raftsmen had to keep arks on the Jersey side of the Delaware and land them opposite New Hope. Wagons would carry the coal across New Jersey.[59] One or two arks that descended the Delaware during November and December may have unloaded their cargoes at Lambertville, New Jersey, and the coal hauled overland to New York.

In March 1815 Jacob Cist tried to get several New Yorkers to handle Lehigh coal on commission or to sell it on their own account. He sent letters to four potential purchasers who might also be willing to act as his agents. Cist offered to deliver anthracite at Lambertville, Trenton, the Delaware tidewater, or New York City. He preferred that they send the names of some people who might be interested in selling anthracite in their city.[60] David Hoosack, one of New York's leading physicians, who had purchased Luzerne County land from Matthias Hollenback, replied that William Dodge was a suitable person to dispose of Lehigh coal. Peace arrived before Cist could market much anthracite in New York City. Nevertheless, he had established contacts for future sales with several of that city's merchants. Meager as the effort appeared, initial contact had been made.

As a partner in the leading mercantile firm of northeastern Pennsylvania, Jacob Cist came into contact with other businessmen in eastern Pennsylvania, Philadelphia, New York, and Baltimore. He knew the needs of the economy and could count on others in helping to market products which would produce a profit. Responding to the fuel crisis in Philadelphia and Baltimore, he sent anthracite from his mines in the Wyoming Valley to both cities. Furthermore, he acquired the aid of other merchants and friends to invest time and money in developing the Lehigh coal trade. Having contacts in all three cities, Cist was able to get a number of merchants and craftsmen to retail Lehigh and Wyoming valley hard coal in their cities. By getting his friends and other businessmen involved in the hard coal trade, Jacob Cist had achieved an important step in the process of marketing Pennsylvania anthracite.

3 Anthracite Advocate

Prior to the War of 1812 few Philadelphia craftsmen had used Pennsylvania anthracite even though many of them knew it existed. The fuel crisis which forced these craftsmen to create the Mutual Assistance Coal Company encouraged Jacob Cist and others to take anthracite to Philadelphia so that the city artisans could experiment with this type of fuel and learn how to use it. As a partner in the Hollenback firm, Cist came into contact with a wide range of craftsmen in the Wyoming and Lehigh valleys who had successfully experimented with hard coal and liked its advantages. Aware of their successful experiments, Cist learned how they used anthracite in their businesses and with the help of Isaac Chapman and Charles Miner used this knowledge to teach Philadelphia craftsmen about burning this fuel. With his family background, business connections, and knowledge of hard coal consumption, Jacob Cist occupied a unique position to help establish a market for anthracite during the war.

Thanks to his father's business, Jacob possessed an intimate knowledge of Philadelphia. Through the efforts of Charles Cist, a number of craftsmen such as Oliver Evans had learned to use and like anthracite. These artisans could be counted on to help Jacob acquire customers. His circle of friends included stockholders in the Lehigh Coal Mine Company who could be called upon for support.

Ultimately, a successful campaign to establish a market depended upon acquiring steady anthracite consumers whose opinions would be respected by fellow craftsmen. Cist had to find Philadelphia artisans in all the trades who used heat for manufacturing purposes and persuade them to give anthracite a fair trial. To win their confidence, he was obliged to understand many crafts and visit numerous shops. Much of his time and effort would be spent getting testimonials and disseminating facts. The information had to show the effect of anthracite on economic activity and convince potential consumers that this fuel was worth trying.

While a number of craftsmen used Cist's anthracite successfully upon its arrival in the city, the glut of anthracite on the Philadelphia market during the fall of 1814 forced the partners into establishing a sales campaign to increase hard coal consumption. Jacob Cist, Isaac Chapman, and Charles Miner spent weeks in the Quaker City advertising anthracite.[1] Before going to Philadelphia, Cist made extensive notes on things he wanted to do and information he wanted to get. His notes indicate careful planning for the campaign. He wanted to see leading craftsmen in every branch of industry that might consume Lehigh coal, and he wrote articles concerning hard coal for newspapers and posted handbills along the streets of Philadelphia.

The outcome of the sales campaign led to the April 1815 publication of a pamphlet written by Jacob Cist titled "On the Importance and Necessity of Introducing Coal, Particularly the Species Known by the Name of Lehigh Coal into Immediate and General Use." Cist had 200 copies printed for distribution.[2] For obvious reasons the pamphlet had a mercantile bias. He wanted to let different types of craftsmen know about the value of anthracite and induce them to purchase his product. Cist's pamphlet was the earliest and most comprehensive statement published in the United States concerning the economic value of anthracite. Men of high reputation such as Oliver Evans and Josiah White, Philadelphia's leading wire manufacturer, gave Cist accurate appraisals of hard coal. Cist's pamphlet indicated the potential impact of anthracite on American economic growth.

His work was also accepted favorably by the public. On 24 March 1815 Charles Miner printed the pamphlet in his Wilkes-Barre newspaper, *The Gleaner.* In this form the pamphlet reached readers along the entire length of the Susquehanna Valley as well as in Philadelphia, Baltimore, and New York City. Six years later Cist sent the booklet to Benjamin Silliman, editor of the *American Journal of Science.* Silliman published the pamphlet as the lead article in volume four of his journal and said that the statements were accurate. Cist's pamphlet underwent at least two more printings. In 1825 and 1827 the Lehigh Coal and Navigation Company borrowed heavily from Cist's pamphlet on anthracite. Parts of the introduction and seven of nine certificates of endorsement reappeared in two Lehigh company pamphlets, one published in Boston and the other in Philadelphia.[3] Consequently, Cist's views received wide circulation.

Foremost on Cist's mind was the introduction of hard coal among Philadelphia's blacksmiths. Over 200 blacksmiths lived in the Quaker City, and most of them used coal in their forges. Cist counted on them to consume the greatest portion of the spring coal shipment. Since almost every smith in the Lehigh, Schuylkill, and Susquehanna valleys burned anthracite, Cist thought blacksmiths in the Quaker City would number among the earliest large-scale consumers. After discovering anthracite's value, they would probably turn to it en masse and help sustain the hard coal trade. Cist planned to consult Thomas Dobson, a master blacksmith and director of the Mutual Assistance Coal Company. He wanted to know how much this blacksmith would pay for Lehigh coal and also wanted him to call a meeting of blacksmiths so that they could be shown how to handle anthracite.[4]

Cist employed a Wilkes-Barre blacksmith to help educate his fellow journeymen in Philadelphia. Chapman and the blacksmith spent three weeks in the Quaker City going from forge to forge trying to attract customers. They offered free instructions for altering equipment and burning hard coal. A bar at the back of the fire needed adjustment, and skill was required to keep the forge open. Since anthracite heated metal much faster than charcoal or bituminous coal, blacksmiths should be careful not to leave iron in the fire too long and scorch it. These artisans had to learn new skills in order to use anthracite in their trade. Changing their habits was not easy.

The attempt to get blacksmiths to use anthracite met with mixed results. After experimentation a few smiths became enthusiastic advocates of the new fuel. For example, Joseph Smith, a Bucks County blacksmith, said that in 1811 he had traveled the road from Fort Allen to Wilkes-Barre via Berwick and found that anthracite was used in every smith shop along the road. He inferred that hard coal must have been valuable since plenty of wood for making charcoal grew near the blacksmith shops. Upon returning home, Smith sent for a wagon load of hard coal. Initial experiments, however, failed to produce results. Lack of knowledge on how to burn anthracite led to its poor reputation among his journeymen. Not until 1814 did one of Smith's journeymen accidentally discover the correct method for burning hard coal. Thereafter he became an energetic supporter of anthracite. He said that Summit Hill coal was the best he had seen. It was so clear of impurities that blacksmiths need not clean out their fire more than once a day, thus saving them valuable time. The Bucks County blacksmith claimed one bushel of Lehigh coal equaled two bushels of Richmond coal or ten of the best charcoal.[5] At that rate anthracite clearly exceeded the value of any other fuel.

Joseph Smith's endorsement went beyond a mere verbal statement. Much to his own advantage and the satisfaction of his journeymen, he used anthracite steadily for thirteen months from February 1814 to March 1815. Smith taught Bucks County and Philadelphia blacksmiths how to use anthracite and secured testimonials for Jacob Cist. A Philadelphia newspaper published one of his statements, and Cist posted another as a handbill. Cist also printed Smith's testimonial as part of his pamphlet on anthracite.

During the war Jacob Cist gained a clientele for hard coal among blacksmiths in Philadelphia. They converted their forges to burn anthracite and learned the essential skills. In April 1815 the Mutual Assistance Coal Company sent out a circular asking a number of questions about the wartime fuel crisis. To the inquiry concerning the use of anthracite, most blacksmiths responded that they had burned every kind of hard coal but preferred softer grades of anthracite or bituminous coal than that coming from Summit Hill.[6]

On 4 April 1815 the Bucks County blacksmith wrote to Cist about his efforts to introduce Lehigh coal. An attempt which he made during March met with little success. His biggest disappointment came when Henry Abbett, a master blacksmith and director of the Mutual Assistance Coal Company, rejected Lehigh coal for Virginia coal in spite of the fact that his journeymen liked to work with Lehigh coal. The journeymen claimed they would rather pay the high price for hard coal than receive Richmond coal free.[7] William Thomas, one of Philadelphia's leading coal merchants, wrote Cist that he wished he had obtained more anthracite the previous fall so that blacksmiths could have become more accustomed to burning it. He said anthracite sales would be slow during 1815 since Philadelphia was once again receiving Virginia coal. If Cist could afford to sell coal at $.40 or $.45 a bushel, the merchant was willing to accept another ark load. He wanted 5 percent commission to sell Cist's coal and would not charge for storage of the coal sold within any reasonable time. During the war Cist had established a market for anthracite among a number of Philadelphia blacksmiths who wanted to continue burning it after 1815.

Jacob Cist extended the campaign to sell anthracite to all areas of iron production. As in the case of blacksmiths, nail makers had burned hard coal in the Wyoming Valley for over two decades. Like Jesse Fell, Cist believed that for nail production anthracite was entitled to a decided preference. He was confident that this branch of the iron industry in the Quaker City would convert to hard coal consumption. In 1810 the city and county of Philadelphia had 20 naileries producing about 3.5 million pounds of nails annually.[8] Cist planned to visit at least seven nail-making shops and from two of these he made the drawing which appears in figure 2. The furnaces had five or nine 0.75-inch bars set 0.75–1 inch apart for heating hoops to cut nails. Both furnaces burned anthracite. This drawing could be used to show manufacturers the nature of anthracite nail furnaces. During his visit to Philadelphia Cist planned to show one nail maker how to use hard coal and hoped to trade anthracite for 10,000 spikes. Another nail-making shop in the old Northern Liberty Market House employed from six to ten men. This shop could become a good customer.

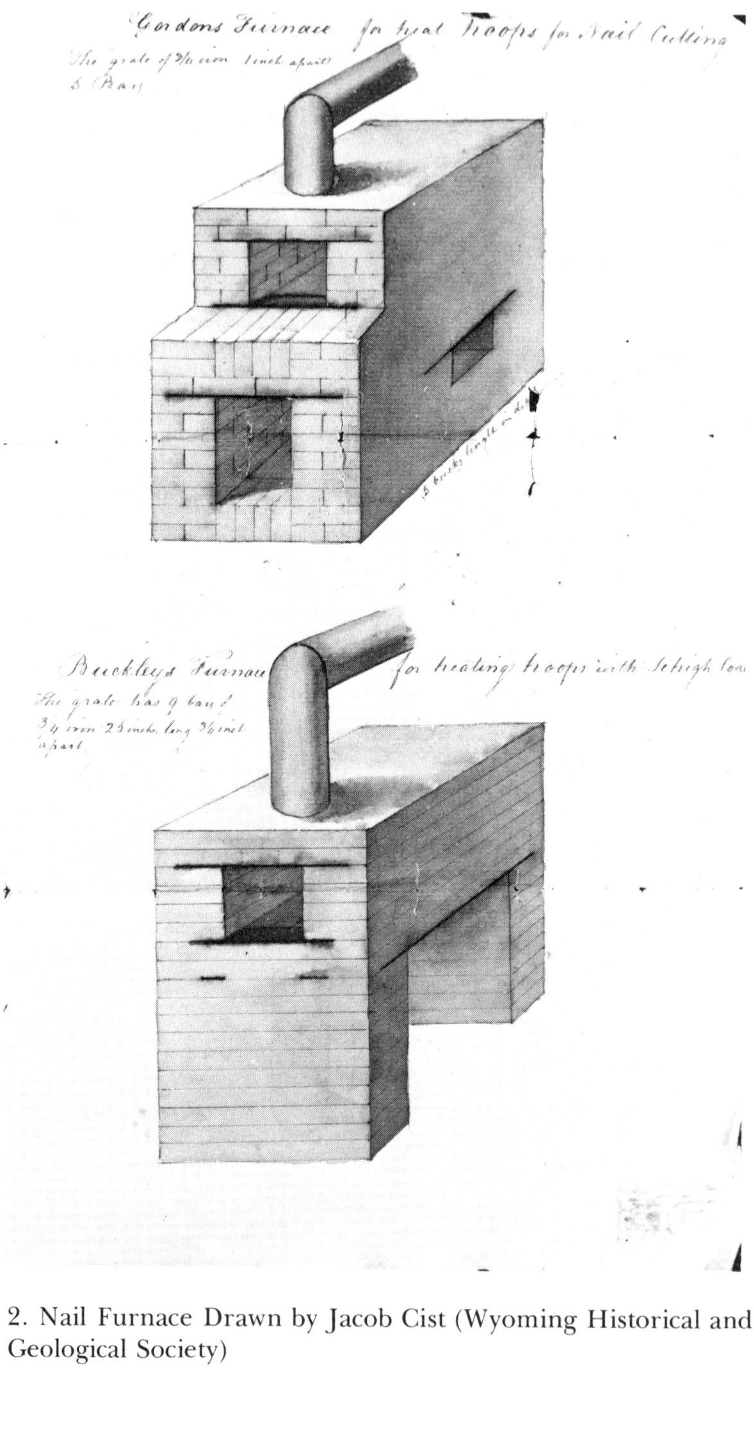

2. Nail Furnace Drawn by Jacob Cist (Wyoming Historical and Geological Society)

Jacob Cist also wanted a testimonial to advertise the excellence of anthracite for producing nails. Cist received a letter of endorsement from three workmen at the Eagle Nail Factory in Wilkes-Barre. They certified that anthracite exceeded all other fuel in value for making nails. Before going to breakfast, they threw hard coal into the fire which made the furnace ready for heating hoops as soon as they finished eating. They liked the advantage of having the fire ready at an early hour in the morning. About half a bushel of anthracite kept the fire burning for 12 hours and lasted until they went home in the evening. They added a small amount of coal at night so the fire would burn until the following morning. At this rate they used less than one ton of coal a month. They also liked the fact that they did not have to pay close attention to getting the fire ready and keeping it burning. Because of this feature an anthracite fire was far less trouble than any other kind of fire.[9]

Furthermore anthracite reduced production costs. The three Wilkes-Barre nail workers heated hoops in half the time and cut a quarter more nails a day with an anthracite fire than with any other type of fuel. Consequently, they got more nails made in the same time period.

Wire production received as much attention as any other branch of the iron industry. Cist's notes called for a visit to the wire factory of Josiah White and Erskine Hazard at the Falls of the Schuylkill. Since they owned the only wireworks near Philadelphia, Jacob wanted to talk with them, show them a Lehigh coal fire, and get a certificate of endorsement. Their rolling furnace was two feet wide, one foot high, and five or six feet long.[10] The fuel crisis had forced White and Hazard to begin looking for a new source of fuel. A Delaware County ironmaster told them he had succeeded in using Lehigh coal in his rolling mill. White and Hazard experimented with two cart loads of Lehigh coal before they accidentally discovered a method for burning it. Their workmen spent an entire night trying to make a fire in the furnace. Giving up, they shut the furnace door and went home. One of the men had left his jacket at the mill and upon returning for the jacket noticed the door was red hot. He opened the door and found the furnace at a glowing white heat. This fire heated and allowed the men to roll four separate parcels of iron. They replenished the furnace with anthracite, let the fire alone, and succeeded in producing a similar quantity of iron.[11] Success in using anthracite made Erskine Hazard and Josiah White strong advocates for the introduction of this new fuel.

White and Hazard testified several times about the value of Pennsylvania anthracite. In 1814 the *American Daily Advertiser* printed one of their endorsements for Schuylkill coal. In November they gave Jacob

Cist a similar statement on Lehigh coal. Seven months later their views again appeared in a report of the Mutual Assistance Coal Company.[12] They liked Lehigh and Schuylkill anthracite for making wire. They believed iron heated with hard coal in their rolling furnace was better than that heated by Virginia coal. These Quaker industrialists considered production costs extremely important. They ran a trial to determine the efficiency of anthracite and bituminous coal. With Lehigh coal three men rolled 0.5 ton of iron for wire in 12 hours. The labor cost $4. In 12 hours the furnace consumed 5 bushels of anthracite. At $.90 a bushel fuel cost $4.50, making production costs with anthracite $17.00 for 1 ton of wire. In a 12-hour day, their furnace consumed 10 bushels of Virginia coal and produced only 0.25 ton of iron. With bituminous coal labor cost twice as much to make the same amount of iron. Assuming the price of Richmond coal to be the same as anthracite, with bituminous coal the fuel bill amounted to four times the cost of hard coal. Comparing the $52.00 per ton of wire produced by bituminous coal with $17.00 per ton made by anthracite, White and Hazard decided in favor of anthracite. They said that Lehigh coal at $.90 a bushel was as cheap as Richmond coal at $.025 per bushel. To them, anthracite made a major difference in conducting business.

The wireworkers also discussed the effect of anthracite on working conditions. For two winters they warmed the mill with hard coal in a cylindrical stove about two feet high and 18 inches in diameter. Their men never complained of ill health or any inconvenience from working around a Lehigh or Schuylkill fire.[13]

White and Hazard came to depend upon hard coal for making wire during the war. If their furnace consumed an average of five bushels of anthracite a day, they needed approximately 50 tons annually. At the beginning of hostilities, they hauled coal by wagon from the head of the Schuylkill River. In 1814 they purchased Cist's Lehigh coal from Steelwagon and Knight. On 6 March 1815 Jacob Cist wrote to them wanting to know how much they would pay for a load of Lehigh coal. White and Hazard replied they would take an ark load at their wharf in Philadelphia for $.50 a bushel. This price, they believed, presented ample inducement for Cist to market more Lehigh anthracite to the Quaker City.[14] White and Hazard wanted to continue getting hard coal.

Farm equipment production was another area that attracted Cist's attention. Jacob wanted to get testimonials from producers of shovels, spades, and ploughs. He received a statement about the role of anthracite for manufacturing ploughs from Joseph Smith, a blacksmith who specialized in producing ploughs at his Tinicum forge. Like Hazard and White, Smith was cost conscious. To forge and weld plough parts together, he calculated that anthracite cost less than charcoal. Much of the hard coal he purchased went toward making ploughs.

Rolling and slitting mills also received Cist's attention. He planned to see a Philadelphia saw maker who was owner or part proprietor of a rolling mill. Knowing that several ironworks in Chester County wanted thousands of bushels of hard coal a year, Cist made notes to contact Jacob Weiss for the names of rolling mill owners in Chester County who would advance cash for 100 tons of anthracite.[15] Of ten or more rolling and slitting mills in counties immediately adjacent to Philadelphia, at least four burned hard coal during the war. In 1812 a Delaware County ironmaster tried to get Lehigh coal for the mill he operated. During the war, however, he depended mainly on Schuylkill County coal for his supply.

On 29 April 1815 the Delaware County ironmaster wrote the Mutual Assistance Coal Company that he had used anthracite at the rolling mill during the war. It heated iron quickly, and the iron remained malleable. Virginia and British bituminous coal opened the pores of iron more and caused greater oxidation if workers were not careful. With bituminous coal the iron would pick up a portion of sulfur. Since anthracite produced none of these adverse effects, he claimed the quality of iron improved with the use of hard coal. Another ironmaster agreed with his appraisal.[16]

Heating iron for rolling mills required about 7.5 bushels of anthracite as compared to 12–14 bushels of bituminous for a 12-hour day. For example, at his mill one Phoenixville ironmaster rolled 2.5 tons a day with 7 bushels of hard coal. In the same time span he used 14 bushels of Virginia coal to make 2 tons of iron. The reason for producing more iron in a day with anthracite, he explained, was that Virginia coal left a greater quantity of cinders. A Virginia coal fire required renewal every time another batch of iron was rolled, thereby necessitating work to stop at least 15 minutes before the oven was fit to receive iron again. With hard coal the oven heated three loads of iron before requiring the removal of ashes to start again. The Delaware County ironmaster said he did one-third more work in a day with anthracite than with bituminous coal. Anthracite burned freely, made an intensely hot fire, and lasted much longer than Virginia coal. Because of its greater heat and durability, he considered hard coal to be worth 50 percent more than soft coal for rolling iron. While these claims may appear extreme, Jacob Cist maintained that for rolling and slitting mills anthracite held a decided advantage.[17] Cist felt that White and Hazard's certificate applied to rolling mills as well as the manufacture of wire.

In New York City the foreman of one of the most extensive iron furnaces in the United States experimented with anthracite in a cupola furnace to produce cast iron. He made four blasts with Schuylkill coal and found it answered all his needs. Pure anthracite without the expense of coking made better iron. The metal was much softer and more suitable for his purposes than cast iron produced with coke. He had no doubt that as soon as anthracite arrived in New York City regularly, it would do away with the use of coke and reduce the cost of casting iron. In 1806 Oliver Evans experimented with Lehigh coal in his cupola at the Mars Iron Works and arrived at the same conclusion.[18] Both Evans and the New Yorker were in contact with Jacob Cist. The New York ironmaster sent him a statement concerning the value of anthracite for casting iron. Cist had two customers who would use Lehigh coal to make cast iron.

Several other branches of the iron industry which could adopt anthracite came to Cist's attention. He wanted a testimonial of its value in a trip hammer forge but got no certificate. When the Mutual Assistance Coal Company asked ironmasters if they had ever used hard coal in such forges, they received negative answers.[19] Cist obtained better results in his attempt to get a statement on anthracite in air furnaces. In 1805 Professor James Woodhouse of the University of Pennsylvania experimented with Lehigh coal in one of Philadelphia's air furnaces. The furnace was filled with half charcoal and half anthracite. Two hours later the charcoal was completely consumed and the furnace was refilled entirely with hard coal. The fire burned for 9.5 hours. Virginia coal, submitted to a similar trial, burned out in 4 hours. Professor Woodhouse said that Lehigh coal promised to be particularly beneficial to industries requiring long continued heat such as for working iron.[20] He sent Cist a statement about the experiment and wished the Wilkes-Barre merchant good luck in his endeavor.

Anthracite also came under consideration for producing steel. Joseph Smith said Lehigh coal worked steel better than any other kind of coal. Likewise, Oliver Evans testified that anthracite was best for making edge tools. Certainly, the growth of the iron industry in eastern Pennsylvania would benefit from the introduction of anthracite. Cist had acquired a number of customers among blacksmiths and ironmasters who would use his hard coal.

Like other anthracite advocates, Jacob Cist became cognizant of hard coal's potential for developing extractive industries. Failure to market Wilkes-Barre coal to a Baltimore glassmaker during the war did not dull his desire to erect an anthracite glass furnace. Cist copied plans for a glass furnace resembling those built in Baltimore, Philadelphia, Pittsburgh, and Washington, D.C. They all consumed bituminous coal. From these plans he designed a furnace which would burn Lehigh coal[21] (figure 3).

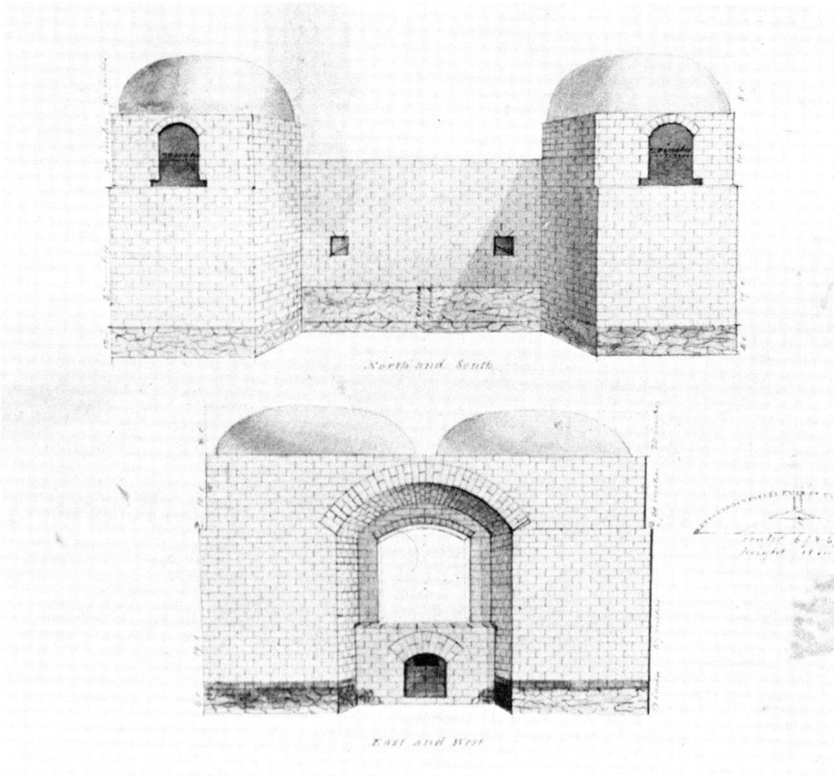

3. Glass Furnace, Washington, D.C., Drawn by Jacob Cist (Courtesy Academy of Natural Sciences of Philadelphia)

Jacob Cist also drew plans for a lead factory and salt works. His design for a salt factory was similar to two establishments manufacturing salt with soft coal. One salt works at Saltsburg on the Conemaugh River ranked among the largest consumers of bituminous coal in western Pennsylvania. If Cist discovered salt deposits in the eastern part of the Commonwealth, anthracite would be valuable in the evaporation process. He may also have envisioned sending hard coal up the Susquehanna River and across the Finger Lakes to the salt works around Syracuse, New York.

Jacob Cist numbered among the early scientists to determine the chemical value of anthracite. He rubbed a bit of anthracite on a fine hard stone in gum water and made the liquid just thick enough to hold the particles in suspension. He hoped this solution would serve as a substitute for India ink. A Luzerne County judge felt that hard coal was superior to lamp or ivory black for printer's ink, paint, and blackening leather. He also maintained it made the best writing ink for records that had ever been discovered. Cist obtained a patent on 28 October 1808 for using anthracite to make this substance. James Woodhouse also made tests on anthracite but came up with no scientific applications. He recommended using hard coal for distilling, making iron and salt, and running steam engines.[22]

In addition to helping create heat for manufacturing, anthracite became a source of energy. Atlantic seacoast cities lacked water power for driving machinery. In Great Britain manufacturers turned to the steam engine for power, and as noted previously, the earliest steam engines in Philadelphia pumped water for the Philadelphia Waterworks. While one trial with anthracite at the waterworks failed, another conducted with Lehigh coal proved successful.

Prior to the war, Oliver Evans also tried anthracite in his steam engine at the Mars Iron Works. He found that hard coal excelled other fuel for keeping a steady fire and running the engine. In addition he discovered that iron heated by anthracite did not scale, as when heated with Virginia coal, and concluded the boiler would last much longer. Moreover, anthracite was smokeless. Soot, being a nonconductor of heat, slowed the process of creating steam. Hard coal left no soot on the boiler and as a result helped the engine run more efficiently. Less anthracite would also produce the same amount of steam as other fuels. Evans believed that Lehigh coal was worth nearly twice as much as Virginia coal.[23]

To consume hard coal, Oliver Evans pointed out, the firebox of a steam engine required special construction. He had altered the steam engine's furnace at the Mars Works to burn Lehigh coal and subscribed for 1,000 bushels annually from Charles Cist. Jacob wrote to Evans about the agreement he made with his father, and asked how anthracite met the needs of a steam engine, hoping Evans would give him a testimonial. Evans answered Cist's inquiry enthusiastically. He liked Lehigh coal and wanted to get more for his business.

On 3 January 1815 Evans told Cist he was constructing a 100-horsepower steam engine for the waterworks. Evans hoped the engine would create demand for Lehigh coal and make Cist's lease a profitable bargain.[24] Cist planned to see the superintendent of the waterworks, who was learning to like Lehigh coal. Steam engines fueled with anthracite would also power Philadelphia's factories. At least one businessman in the Lehigh Valley considered adopting steam to run his mill. On 11 February 1815, John Cunow wrote to Cist that he still planned to build a steam gristmill. He thanked the Wilkes-Barre merchant for sending him plans of one of Oliver Evans's gristmills that could be run with a steam engine. If he constructed the mill, the Bethlehem resident said he would purchase anthracite from the Lehigh mine.

As in the case of ironmasters, Jacob Cist had a market for his hard coal among steam-engine operators. He wanted to use hard coal as fuel for steam-powered vehicles. In 1810 Isaac Chapman had written to the American Philosophical Society about the possibility of using an anthracite-burning steamboat on the Susquehanna River. He sent his plans to the society for examination and evaluation. Robert Patterson, the president of the society, prepared a careful statement covering the details of Chapman's invention, arguing that because of the shallow bottom, rapid current, and numerous rocks in the Susquehanna River, Chapman's steamboat would probably not successfully navigate the river.[25]

Cist accepted Patterson's opinion concerning Chapman's boat. On the other hand, he hoped that anthracite could be used successfully on the type of steamboats invented by Robert Fulton, Oliver Evans, and James Fitch. He planned to go aboard a steamboat at Camden and examine it carefully. He also asked Evans to send an estimate of expense for one of his steam wagons. Cist wanted to put 20–25 steam wagons, carrying 1.5 tons of coal, on a railroad with a descent of about ⅓ inch per yard. He asked whether Evans's steam wagon was similar in principle to one which had appeared in a recent London paper. Evans replied that he had been declaring the practicality of steam carriages for nearly 40 years. He was not sure whether a steam wagon similar to his had been developed in England. Evans could not state the expense for a carriage since he had not perfected the plans. He wanted to get aid from Congress to develop a steam railroad and said he would invest $10,000 to test the value of his idea.

Jacob Cist summarized the value of hard coal for manufacturing purposes in a few sentences. The great value of anthracite was its durability and regularity of heat. Hard coal lasted twice as long as Virginia coal. The greatest advantage of this fuel was the increased quantity of work which it enabled men to do on account of the quickness of heat and the little attention required by the fire. Cist rated anthracite three to one over Virginia coal and for some purposes as high as twenty to one. He also paid attention to its effect on working conditions. Rooms remained free from smoke or smell. Thus use of hard coal did away with the need for contrivances for consuming smoke. Without such contrivances, factories were nuisances creating air pollution in their immediate neighborhood. Like Cist, a few Philadelphia craftsmen and industrialists recognized these facts. Pennsylvania anthracite would provide the heat for many craftsmen and make the city a clean place in which to work. These points converted a number of artisans into anthracite customers.

While Cist was primarily interested in establishing a market for anthracite in Philadelphia, he also wanted to win friends for anthracite among the agrarian population. He planned to see James Mease, a director of the Philadelphia Agricultural Society, about getting support for the introduction of Lehigh coal. Evidently, the society was sympathetic to proposals which would increase farm productivity. In 1813 a member of the society read a paper on the use of lime in Great Britain. To make fields more productive, the English adopted fertilizers such as lime. Coal burned a great deal of lime to fertilize British farms so that the land would yield greater crops. While growing up in Craven, England, one member of the society saw farmers using lime to enrich their grazing land. Before the introduction of lime, Yorkshire land was scarcely worth cultivating. Lime provided fertilizer for Rockingham's turnips, thereby stimulating agrarian production.[26] Since Pennsylvania had an abundance of lime and coal in close proximity to each other, the same process could take place in the Keystone State. Anthracite could become the fuel for making lime to fertilize fields in central and southeastern Pennsylvania.

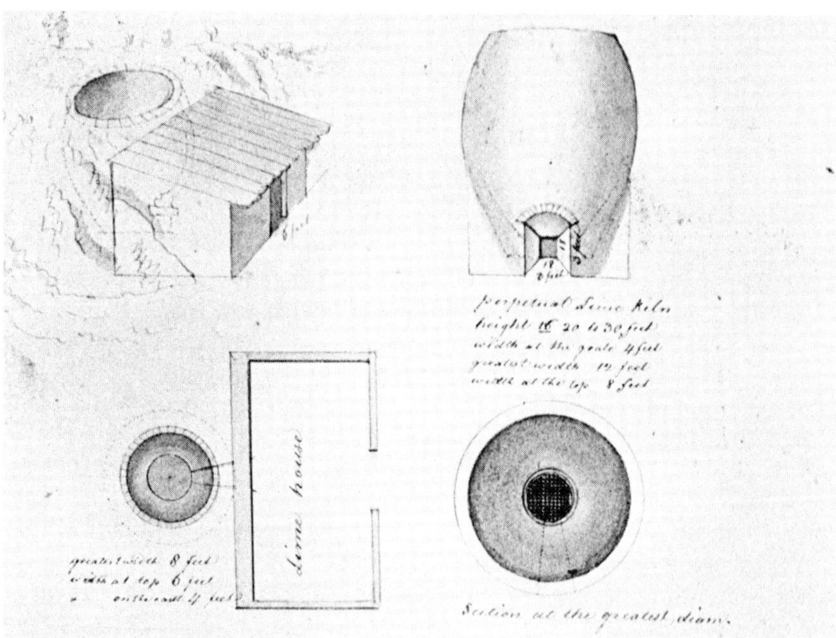

4. Plan of Lime Kiln by Jacob Cist (Courtesy Academy of Natural Sciences of Philadelphia)

Jacob Cist considered using hard coal for producing fertilizers. He made drawings for limekilns (figure 4) and a potash furnace. In 1810 southeastern Pennsylvania had about 450 kilns, which burned nearly one million bushels of lime annually. Lancaster County possessed the only potash furnaces east of the mountains.[27] These extractive industries could burn anthracite, and some in the Susquehanna Valley probably did. Cist could supply fuel to lime and potash makers in the Delaware, Lehigh, and Susquehanna valleys.

Wheat was Pennsylvania's biggest agricultural product. Before the building of canals, the greatest grain-growing area in the United States centered in central and southeastern Pennsylvania. Most of the grain was ground into flour or turned into beer and liquor. Oliver Evans's inventions revolutionized the milling process. As noted previously, Jacob Cist hoped to sell Lehigh coal to at least one steam gristmill constructed upon an Evans patent. Success could mean expanding sales to other steam gristmills.

In addition to selling fuel to steam mills, Cist planned to market anthracite to breweries and distilleries. According to the federal manufacturing census, seven counties in central and southeastern Pennsylvania had 22 breweries, and 11 more were located in the Quaker City. Within this same region there were nearly 1,500 stills.[28] Cist's notes contain the names of a dozen distillers with whom he wanted to talk. In Philadelphia he felt that William Lattimore and another distiller would use Lehigh coal. The greatest number, however, lived in the Lehigh Valley. For anyone wishing to burn Lehigh coal, Cist drew the picture in figure 5. The grate should be composed of eight loose one-inch-square iron bars split at each end. These bars should be placed from ⅝ to ¾ inches apart. The bottom of the still should be 18 inches above the grate. Cist believed this was the proper way to erect a grate for distilling liquor. He also wanted to see the breweries of Philadelphia.

Jacob Cist wanted certificates which showed anthracite's excellence for distilling and brewing. William Brown, a Philadelphia brewer and distiller, testified on 10 December 1814 that he had used Lehigh coal during the previous two years. Like other businessmen accustomed to the qualities of hard coal, he preferred it to wood. Lehigh coal was cheaper, safer, and required less labor. For distilling 100 bushels of grain, he used one ton of hard coal and one cord of wood. To dampen the fire while mashing or drawing off the still, he threw on some coal. A stick or two of wood on the fire quickly raised the heat to the proper temperature again. The process required ten days. At that rate his still consumed about 35 tons of anthracite annually. An Easton distiller gave Jacob Cist a statement repeating points made by the Philadel-

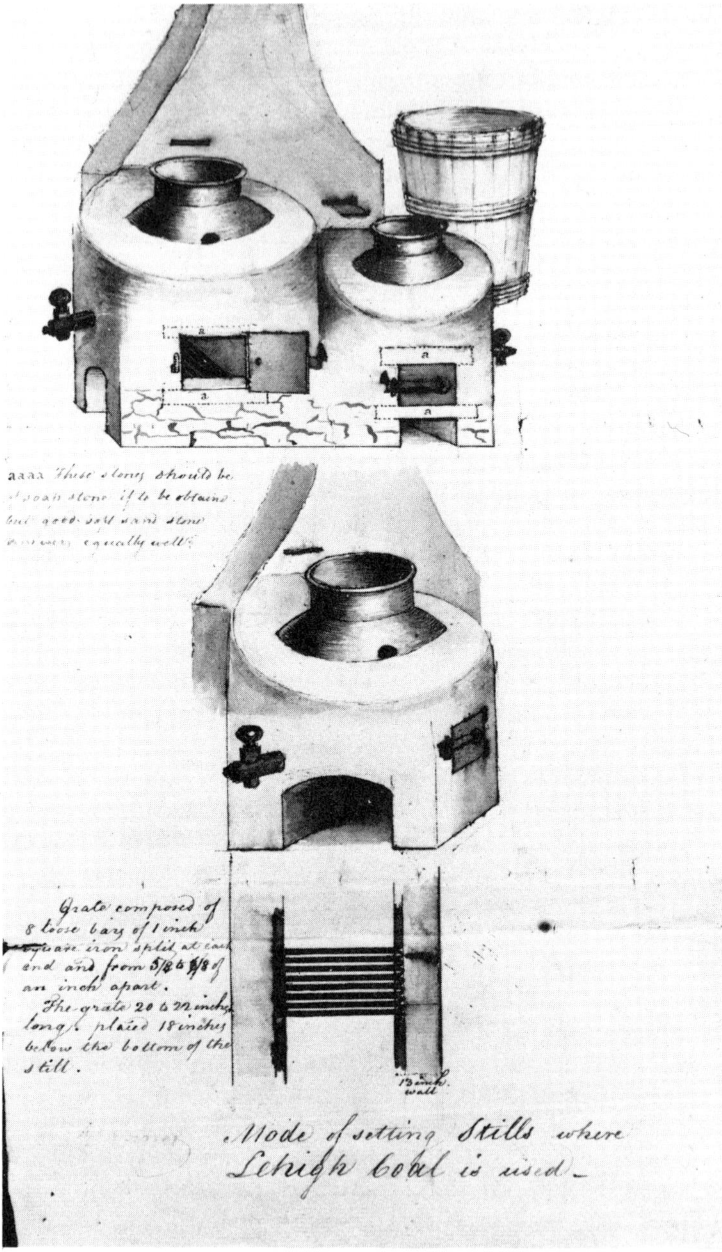

5. Still Drawn by Jacob Cist (Wyoming Historical and Geological Society)

phian. He too noted the fine qualities of the fuel which kept his fire going for two previous years.

One other consideration concerning the effect of anthracite on agriculture entered Cist's mind. In the preface to his pamphlet, Jacob said that by using hard coal farmers no longer needed to retain large acreage in woodland. The long hard hours consumed in cutting firewood for home use could be eliminated, or farmers could use this time and energy to cut firewood they could sell in nearby towns and cities. On the other hand, Cist pointed out that in the agricultural counties of Lancaster, Dauphin, and York hard coal provided the fuel for blacksmiths and farmhouses. In 1814 these counties consumed about 2,000 tons of Wyoming Valley coal. Anthracite replaced the wood which came from local farms. Instead of supplying homes and smiths with fuel, farmers could reduce the size of their wood lots to a point sufficient for constructing buildings and fences. As a result, the quantity of usable land could be increased without any additional purchase, and superfluous wood could be sold to advantage.[29]

Depletion of woodland near American cities worried Jacob Cist. He wrote to one editor that people not in the habit of reflecting or taking a retrospective as well as a future view of things could hardly imagine how much the price of fuel and woodland had risen in recent years. The inhabitants of many villages knew that only a few years ago they paid practically nothing for firewood and now paid $4, $5, or $6 a cord. If they looked forward, they might feel that in only a few more years they would pay $7, $8, or $10 a cord.[30]

Cist's comments were also applicable to the Quaker City. In *The Picture of Philadelphia,* written on the eve of the War of 1812, James Mease referred to the city's fuel situation. He said that homeowners in Philadelphia still relied heavily upon wood for heating and cooking. According to Mease, from 1808 to 1810 more than 65,000 cords of wood had landed annually in Philadelphia. He felt that an additional two-thirds of that amount had been landed at public wharves in Southwark, the Northern Liberties, and along the Schuylkill waterfront. Like previous city leaders, he commented about the increasing cost of firewood and looked forward to the day when an inexpensive and abundant amount of Pennsylvania anthracite would come to the Quaker City. After noting the use of coal by many craftsmen in Philadelphia, he said that coal would be used to heat counting houses and public buildings and offered the potential of greatly reducing fuel costs of homeowners. If coal was used extensively by the city, he noted, the price of fire insurance policies would drop sharply.[31]

Like James Mease, Jacob Cist thought about the amazing consumption of wood by numerous craftsmen, industries, private homes, and business establishments. He calculated a glassworks required from 1,000 to 12,000 cords of wood a year and ironworks needed 5,000 or 6,000 acres of woodland. Other industries demanded equally large amounts. The time had arrived, he believed, to substitute anthracite for firewood.

Cist paid attention to requirements for making charcoal. Not every kind of wood answered the purpose. Charcoal makers preferred hickory, maple, gum, and chestnut. To cut trees and stack logs on piles required time. Four cords of wood produced about 100 bushels of charcoal. Much waste and destruction took place preparing logs. Even in spite of every care, a pile of charcoal could be destroyed in the burning process. Watching the burning pile carefully night and day took still more energy. In addition to pointing out the difficulty in preparing charcoal, Cist stated that as much work could be accomplished with one bushel of hard coal as with eight or ten bushels of wood coal. Furthermore, time would be saved. Cist asserted that use of Lehigh coal would soon supersede the use of charcoal altogether.[32]

On 22 December 1815 James Ronaldson computed the cost of heating Philadelphia with wood fires. He told Cist that oak had increased to $9 a cord and hickory sold for $12 a cord. Assuming the population of Philadelphia to be 100,000, with six people in a family, he rounded off the figure to 16,000 households. He believed that each family consumed ten cords of wood annually; at an average price of $7 a cord, the population paid approximately $1,112,000 for fuel to heat their homes. Bakers, brewers, brick burners, distillers, and potters consumed about 50,000 cords or nearly $750,000 worth of wood, making the total fuel bill for Philadelphia about $2 million a year. If Cist could meet the fuel demands of the Quaker City with Lehigh coal, the public might save 25 percent on its annual fuel bill.

Cist knew the population had to be convinced that anthracite would burn without difficulty in a grate or stove for domestic purposes before they would convert from wood to coal. Furthermore, he had to prove that Lehigh coal was safer and cheaper than wood. To achieve this goal, Cist designed the stoves which appear in figures 6 and 7. They consumed about one peck of coal a day to warm large rooms or cook food. A forge near Bethlehem manufactured these stoves for Cist.

Jacob wanted an ironmaster to send 4 or 5 of his stoves to Philadelphia so that Lehigh coal could be given a fair trial.[33] Cist hoped anthracite would meet with the same approval for heating public buildings as it had for warming shops. Cist planned to have anthracite tried in Philadelphia at a physician's office, a mansion, a bank, and the jail. He also believed Charles Willson Peale would use Lehigh coal to heat the Philadelphia Museum.

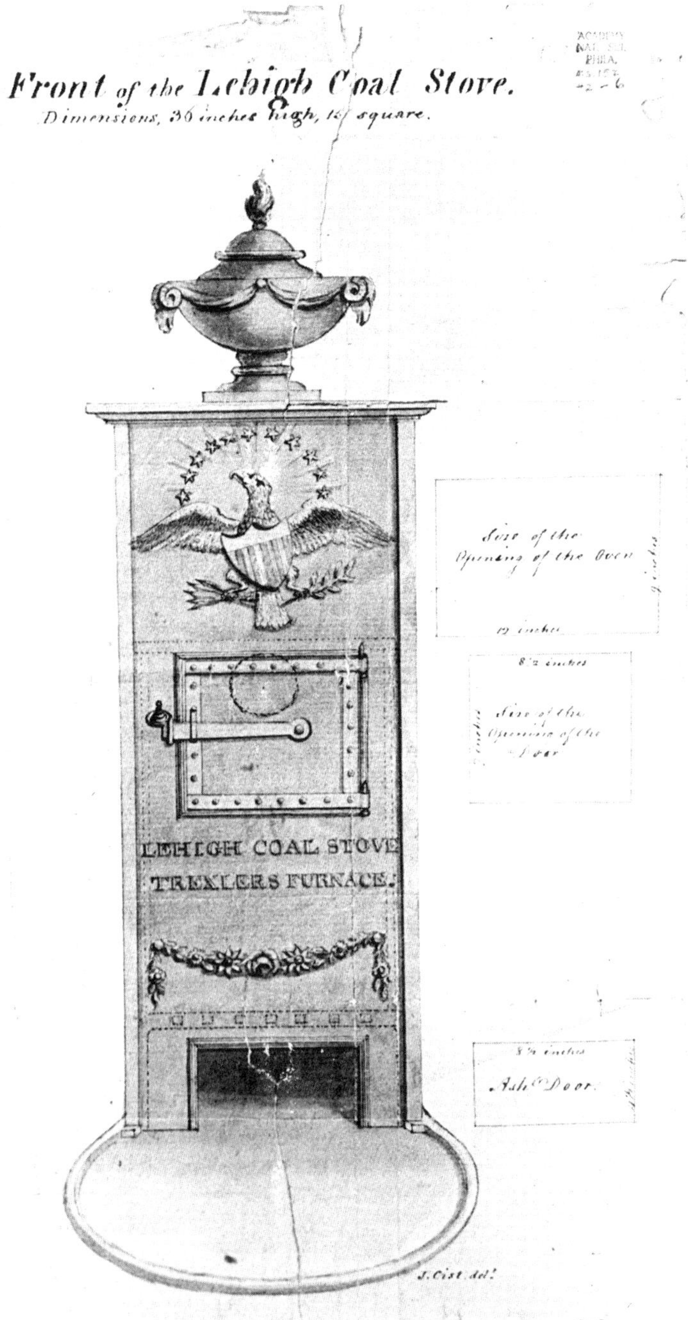

6. Lehigh Coal Heating Stove Drawn by Jacob Cist (Courtesy Academy of Natural Sciences of Philadelphia)

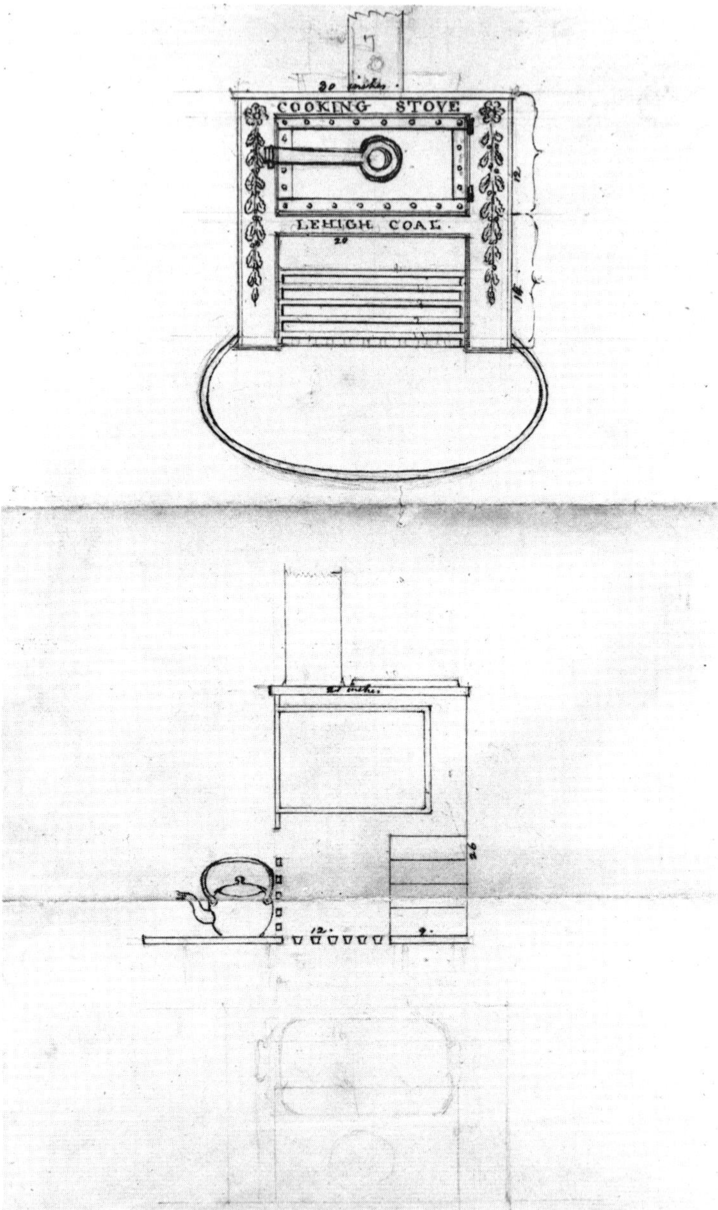

7. Lehigh Coal Cooking Stove Drawn by Jacob Cist (Wyoming Historical and Geological Society)

The cashier of the Pennsylvania Bank tried Lehigh coal in a large stove and found that it met his need exceedingly well. He gave Cist a testimonial which repeated attributes mentioned previously. Anthracite produced excellent heat and burned lively. He believed that Lehigh coal lasted twice as long as any other coal brought to market, and it required less labor and attention. The cashier also liked the clean burning fire and observed no smoke, smell of sulfur, or flying dirt when stirring the fire. These qualities were of extreme importance if Cist wanted anthracite to be accepted for family use.[34]

In several articles submitted to newspapers, Jacob Cist added more information about the value of anthracite for domestic use. He said that 1.5–1.75 tons of Lehigh coal would amply heat a large room for one winter. Homeowners would obtain more warmth and comfort from that quantity of hard coal in one of Cist's square stoves than from six or seven cords of wood. The fire in these stoves seldom went out and required little attention, whereas wood fires demanded constant attention and required at least an hour to begin warming a room. With hard coal rooms stayed at relatively constant temperatures. Security from fire was another important feature since city dwellers feared the devastation caused by fire. With the adoption of anthracite, the chance of stovepipes and chimneys catching fire decreased.

Several Philadelphians liked Lehigh coal for heating their homes. Thomas C. James, a leading Philadelphia physician and stockholder of the Lehigh Coal Mine Company, started burning anthracite at his home during the winter of 1804 and continued using it thereafter. In order to heat his parlor during the winter of 1814–1815, Josiah White used hard coal in a grate set on andirons in an open stove.[35] Both consumers preferred anthracite to other fuel. On 27 October 1814 the stockholder wrote to Cist that a steady supply of Lehigh coal would be a great acquisition for the Quaker City. Nevertheless, Philadelphia still had to be taught the proper mode of burning it. He hoped that Cist's stoves would permit hard coal to start burning easily. Otherwise, due to its comparative difficulty of ignition, anthracite would be considered insufficiently flammable for domestic purposes. Five months later White and Hazard wrote Cist about the domestic consumption of anthracite. If Lehigh coal could be sold in Philadelphia for $.50 a bushel, families would purchase it to a considerable extent. At that price, they felt families would purchase 50 bushels annually, thereby enabling Cist to dispose of a very large quantity.[36] Counting on an average of 1.67 tons for each family, Philadelphia's 16,000 households would need nearly 27,000 tons of anthracite a year.

The overall impact of anthracite on the urban environment also caught Cist's attention. As had often been observed by travelers, Cist noted, London was constantly enveloped in a cloud of dull smoke, especially during wintertime. To early morning travelers approaching London from adjacent heights, this smoke appeared to threaten the city with an impenetrable darkness. If curious visitors wished to get a bird's eye view of London from the Tower of London, the dome of St. Paul's, or from any other towering structure, they had to select a favorable time when there was little smoke in the air or when winds drove the smoke away quickly.[37] Like others, Cist hated the soft coal smoke which polluted London air. He believed that anthracite would help keep Philadelphia and other American cities clean.

Cist also foresaw social benefits which would come from urban consumption of hard coal. Anthracite would provide the long desired and only practical remedy for a major urban problem. This type of fuel totally eliminated work for a wretched class of unfortunates, the chimney sweepers. These poor people could be employed in more productive work.

Jacob Cist did not see Pennsylvania anthracite as a completely new fuel for Philadelphia since he recognized the similarity between Lehigh coal and that produced in Kilkenny County, Ireland. In 1802 the Dublin Society published a book titled *Statistical Observations Relative to the County of Kilkenny*. It noted the excellent quality of Kilkenny coal, which contained 97.3 percent pure carbon and provided fuel for homes and industry in Cork, Dublin, Londonderry, and Waterford.[38] Since the 1790s Kilkenny coal came to Philadelphia as ballast and furnished the Quaker City with fuel. Businessmen, like Cist, could easily compare Kilkenny coal with the substance obtained from Penn's woods. To break down prejudice against Lehigh coal, Cist pointed to this Irish fuel and its benefit to the economy. The similarity between Pennsylvania anthracite and Kilkenny coal meant that Philadelphians had consumed hard coal for at least two decades. Consequently, they need not fear experimenting with a brand new fuel.

The lesson of Ireland's sending coal to the United States did not go unnoticed by Cist. He envisioned shipping anthracite to Russia and wrote to a fellow merchant that hard coal might become an important export from the United States and would help expand commercial relations between the two countries. Anthracite could be delivered cheaper on the Atlantic seaboard from mines located in northeastern Pennsylvania than from any other point in the country. Hard coal mines were of great extent, some strata being more than 20 feet in thickness. Such mines could supply an abundance of coal for international trade. Cist proposed establishing an export business in anthracite and felt that the Russians would especially want to acquire Pennsylvania anthracite.[39] Implicit in Cist's thinking was the desire to number among the nation's mercantile princes. He hoped to supply eastern Pennsylvania, the Atlantic seaboard states, and a European nation with anthracite.

During the War of 1812 Jacob Cist ranked as the foremost anthracite advocate. He considered the potential impact of Lehigh coal on almost every branch of industry mentioned in the 1810 federal manufacturing census and used his influence to establish a market for it among a number of craftsmen and industrialists. He envisioned anthracite as a prime mover in the American Industrial Revolution. Agriculture as well as industry benefited from burning hard coal. Without anthracite Atlantic seaboard cities would not have developed to their full potential.

4 Completing Plans for the Lehigh Coal Trade

After establishing a market for Pennsylvania anthracite among Philadelphia's craftsmen and firmly believing that the rest of the city would become hard coal consumers once an abundant supply could reach the city at a low price, Jacob Cist turned his attention toward completing plans for the Lehigh coal trade. This phase of the marketing process dealt with capital formation—the investment of managerial talent and money into creating mining and transporting facilities to get anthracite to the Quaker City. As far as the Lehigh coal trade was concerned, this phase fell into two time periods. During the first stage from 1814 to 1818, Jacob Cist and Isaac Chapman shared the major responsibility for capital formation. In 1818 Josiah White, Erskine Hazard, and George F. A. Hauto, a German mining expert, took over the management from Cist and Chapman. With the help of Philadelphia financiers they ultimately created the Lehigh Coal and Navigation Company. Although White and Hazard deserve most of the credit for this achievement, the role of Cist and Chapman cannot be overlooked.

On 7 April 1815, nearly six weeks after news reached the United States that a peace treaty had been signed with Great Britain, Cist wrote to Chapman that he saw no other business opportunity which promised to achieve such magnitude and make more profit than the anthracite trade. The price of hard coal still remained fairly high in Philadelphia, and he had acquired a number of customers who wanted to receive a steady supply of this fuel.[1] Substantial capital improvements in the Lehigh coal trade due to the fuel crisis also forced Cist to continue his efforts. He did not want his money and energy wasted at a time when other businessmen showed an interest in his project.

Jacob Cist's ambition to introduce Lehigh coal went beyond the desire to make quick profits from wartime conditions. Cist clearly intended to establish the Lehigh coal trade on a permanent basis. He knew that Philadelphia needed a good fuel supply and believed that Lehigh coal would meet this demand. The anthracite trade would make money.

To insure control of the Lehigh Coal Mine Company, Cist began purchasing its stock during the fuel crisis. A list of stockholders in 1814 accounted for 39 of the original 50 shares. Of the 29 people on the list, Cist noted who owned shares outright and who laid claim to others. His mother had four shares and the company president owned three. Everyone else held three or less. Among the names on the list, a Bethlehem merchant owned three, John Nicholson's estate had two, Oliver Evans claimed two, one of the original Nazareth stockholders still held two, another original Philadelphia stockholder had two, an executor of a Philadelphia estate controlled one, and Tench Coxe was not paid up.[2] On 27 November 1814 Cist purchased the two shares from the Nazareth stockholder for the sum of $300. One month later Cist gave the Bethlehem merchant a note of $600 for his three shares.[3] Counting the four shares which he owned before these transactions, by the end of 1814 Jacob Cist had become the largest stockholder in the mining company. Jacob Weiss was the only other major shareholder. While Cist assumed the lead in promoting the company, Weiss and his family gave Cist strong support.

Cist relied on his partner, Isaac Chapman, to superintend the enterprise. Cist, Chapman, and Charles Miner became the main backers of the Lehigh coal trade from 1815 to 1818. The policy of making leases with specific individuals for certain tasks continued as it had during the war. Cist assumed the responsibility of supplying coal and left the job of getting it to market up to his partners.

During the fuel crisis Cist started the process of capital formation. Each step, from mining coal to its final distribution in Philadelphia, required careful planning as well as the investment of time and money. As noted previously, one of Cist's major responsibilities involved mining and getting anthracite from Summit Hill to the Lehigh River. This responsibility led him to make one of the first sketches of the geological formation at Summit Hill.[4]

Following the British techniques introduced by the first superintendent of the Lehigh Coal Mine Company, Cist drew three major coal seams which outcropped at Summit Hill. One was 9 feet thick, a second 50 feet thick, and a third averaged 10–15 feet. All three were for good open quarry mining and attracted the attention of merchants and financiers interested in developing the Lehigh coal trade. Cist's drawing was a prerequisite to capital formation.

Removing the trees, rocks, and other debris from atop the coal in 1813 and 1814 was part of capital formation. As noted previously, this task gave Cist and his partners considerable trouble. By 1815 they had opened the quarry in order to mine on a fairly large scale.

Although Cist's long-range plans called for constructing a railway, the dirt roads which Chapman and others laid out from Summit Hill and Nesquehoning to Lausanne and Mauch Chunk must be considered as capital improvements. Exploring the terrain for the best descending routes from the mines to the river required time. Removing trees and rocks from these paths added to the operating costs, but these paths later served as roadbeds for railways.

Building facilities to construct arks also took time and must be considered as part of the capital improvements. In addition to acquiring rights to the sawmill, on 9 March 1814, Isaac Chapman started laying out an ark yard at the mouth of Nesquehoning Creek. The ark yard became an important part of the operation as a place for building boats to take coal down the Lehigh River.

A railroad was not the only permanent improvement Cist and Chapman planned. In 1811 the Lehigh Coal Mine Company had leased to a local resident a house in Lausanne and a sawmill at the mouth of Nesquehoning Creek. In August 1814 Cist acquired the lease, allowing the resident the exclusive use of the barroom and a small room on the second floor. Cist and the occupant had joint use of the garret, cellar, and stoop. If the occupant wanted to saw lumber at the mill, he retained that privilege as long as he did not interfere with Cist's business and paid a fee of $.25 for each day's use.[5] Under such arrangements this man helped Cist in the coal business. During the coal mining and rafting season, his place became a boardinghouse.

On 26 December 1814 Isaac Chapman wrote Cist about plans to build a second house at Lausanne. Chapman wanted a full-time carpenter at Lausanne to operate the sawmill and to build boats and houses. Without another permanent residence for a family, it was impossible to attract such a craftsman. Chapman planned to construct a house with a store in one room, space for a family in another, and sleeping rooms upstairs. He asked Cist to make arrangements with the company so that the house could be built on company land.[6]

Chapman suggested another idea which he felt would be a better way of establishing a settlement at Lausanne. He wanted the Lehigh Coal Mine Company to survey a village plot and sell a certain number of lots to anyone who would build on them. If a number of mechanics decided to build there, the community would eventually benefit from their presence. Cist liked Chapman's suggestion and drew the plan for the village of Coalville which appears in figure 8. Accompanying the drawing was the following description of the town, dated Wilkes-Barre, 14 March 1815.

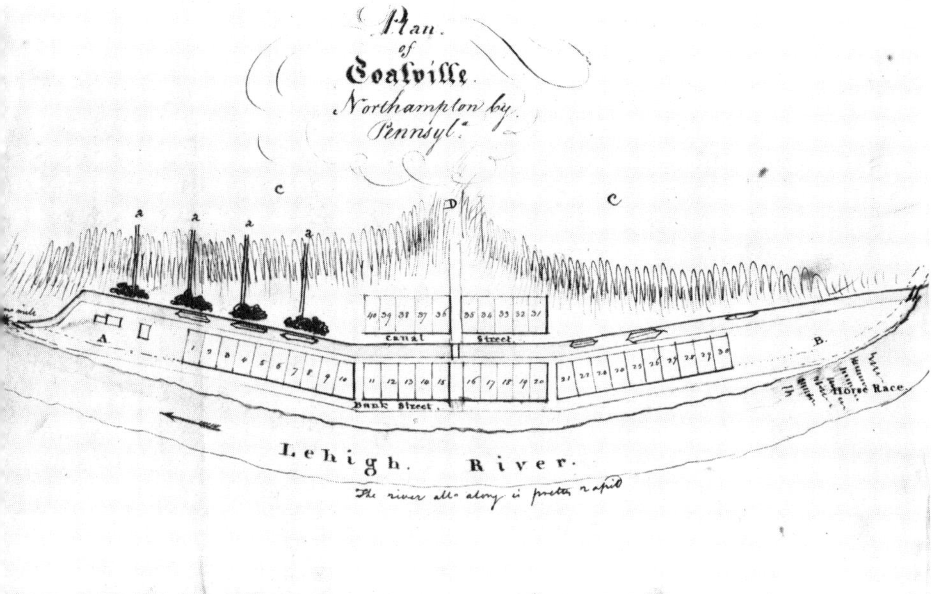

8. Plan of Coalville (Mauch Chunk), 1815, by Jacob Cist (Wyoming Historical and Geological Society)

I send as I promised you a sketch or eye view draft (it shall be surveyed when the snow is off the ground) of the small flat on the Lehigh which would admit of a small village of sufficient size to accommodate a few mechanics, viz., a blacksmith, wheelwright, shoemaker, carpenter & together with boatmen—all which it is indispensable to have on the ground and that very soon if the company ever intend to have their mines worked to advantage & with economy [*sic*] & in a regular systematic [*sic*]. The canal will be necessary in case the Lehigh is ever improved to enable ascending loaded boats to pass the "Horse Race". It is already made being the race dug . . . by McMurray for his sawmill. When finished, it will afford a complete harbor where boats and arks will be protected on the breaking up of the ice and can lay ready loaded at all seasons out of the reach of accidents occasioned by sudden rises of the river.

The lower (A) & upper (B) parts of this flat ought to be reserved for improvements or mills hereafter to be made by the company, that is the water power or mill seats at the Horse Race and at the lower part of the canal. The Nesquehony Hill (C) breaks off very abrupt on to the flat. It would be desirable to have the waggons discharge their loads at the top of the hill & to deliver the coal from "staiths" a a a a or through at the canal.

> [At] A & D there is a hallow where teams can get up, but it would be better that the company when they once work these mines, should have their stabling on the top of the hill.
>
> The competition which will now be experienced from the Virginia coal has alarmed my associates and I shall not be surprised if they fly the ground. Indeed I expect it, but I shall not despair if the company only sanction the arrangements necessary to be made, of rendering this property valuable to them as well as myself both as a lessee & stockholder.[7]

Instead of becoming known as Coalville, the location described by Cist took on the name of the small creek running down the ravine. During the spring of 1815 people started calling this Lehigh River coal landing Mauch Chunk. The nucleus of a village to service the coal trade began to emerge (figure 9). Like the work at the mines and on the roads, construction of these buildings also counted as capital improvements. Some of the work was paid for by bartering goods and services for anthracite, but Cist paid for most of the improvements in cash.

9. Mauch Chunk, in Eli Bowen's *The Pictorial Sketch-Book of Pennsylvania* (Philadelphia: W. White Smith, 1854)

When the War of 1812 ended, the Lehigh coal trade fell on hard times even though Cist and Chapman continued to send some anthracite down the Lehigh Valley each year until the enterprise was taken over by Josiah White and Erskine Hazard. During 1815 Isaac Chapman engaged the services of several Lehigh Valley residents to help with the coal business. He was extremely pleased in having secured a carpenter to take up residence in Mauch Chunk. Chapman described him as a hard-working man with a small family who had been acquainted with the lumber business for several years and was willing to carry on such a business in Mauch Chunk. The carpenter had agreed to supply in advance his own provisions for a year, one or two cows, and an oxen team. He promised to aid in meeting expenses for buying timber or timberland and would receive a portion of the lumber as compensation. In order to obtain the necessary material for building arks and boats at the least possible expense, Chapman agreed to become the carpenter's partner in the lumber business. When his team was not engaged in getting lumber, it would haul coal. The carpenter agreed to help Chapman by furnishing men to cut trees and get timber to the sawmill.[8] With his aid the coal trade could be carried on more efficiently. Other craftsmen were also willing to move to Mauch Chunk and help establish the coal trade on a permanent basis.

In order to determine the demand for anthracite in Philadelphia, at the beginning of March 1815 Jacob Cist wrote to two anthracite customers and two retail coal merchants asking if they could use more Lehigh coal. On 11 March William Thomas, one of the coal merchants, replied that Cist's coal would not sell as readily as it had during the war due to the competition from Virginia coal. He was willing to try an ark load as long as Cist sent the lowest priced coal. He said that his commission would be 5 percent, but he would not charge for storage if the anthracite sold within a reasonable amount of time. He predicted that coal prices would soon drop from $1.00 a bushel to $.40 or $.45 per bushel.[9]

James Ronaldson informed Cist on 13 March that he would take 100 or 200 bushels of Lehigh coal if the price compared favorably with other coal. He gave Cist permission to extend the quantity to an ark load as long as the vessel did not carry too many tons. Ronaldson's workmen had developed a decided preference for hard coal during the war, and he told Cist that in the city as well as the country the interest of a manufacturer depended upon pleasing his journeymen.

One of Cist's biggest customers during the war responded on 18 March. Josiah White and Erskine Hazard said they would pay $.52 a bushel for an ark load of Lehigh coal to be delivered as early as possible at their wire factory. If the coal was delivered at their wharf in Philadelphia, they agreed to pay $.50 a bushel within 90 days. This price, they felt, provided ample inducement to convey coal to Philadelphia. At $.50 a bushel they believed it would be used by families to a considerable extent. If Cist sold a softer grade of coal than that retailed by Steelwagon and Knight, the wire manufacturers preferred it.[10] Of all the responses, White and Hazard's showed the strongest desire to acquire anthracite for their factory.

Even before receiving this information Jacob Cist and Charles Miner decided to continue mining Lehigh coal. On 17 March 1815 they placed an advertisement in Miner's Wilkes-Barre newspaper for coal diggers. They advertised for 700 tons of anthracite to be dug at the Summit Hill mine. The coal had to be of a convenient size for hauling and would be weighed at the river.[11]

Just how much coal was dug at the great coal bed during March and April 1815 is impossible to determine. On 6 April Isaac Chapman said he had no assurance of getting more than 40 tons during the season.[12] Receipts indicate that at least this amount was hauled to the landing. Although this fell far short of Cist's expectation, the total amount hauled was sufficient for several ark loads.

Isaac Chapman took charge of constructing arks during 1815. He began operations at Lausanne on 4 February, purchased one ark bottom in March, and took it to Weiss's landing. He had difficulty, however, in getting labor to work at the ark yard. Cist met with similar results in trying to get carpenters in the Susquehanna Valley. Finally, at the beginning of May, Chapman succeeded in getting two carpenters to build boats, but by this time it was too late in the season to take advantage of good rafting conditions.[13] On 27 May they had finished four boats and fitted them for running coal. Chapman had enough boats to ship his coal.

The arrival of peace made it difficult to raise funds for carrying on the coal trade. On 31 March Chapman wrote Cist that peace in the public opinion had such an effect upon the trade that he found it much harder to raise money than during the war. Chapman also preferred to learn by experiment the effect peace would have upon their business. To determine the demand, during the approaching season he planned to ship about 140 tons, if he could acquire that amount of coal. The quantity which he might want after another winter had passed would rest on subsequent events and primarily on his success in selling coal.[14]

While lumber descended the Lehigh to Easton during the last week in March 1815, Chapman was in no hurry to ship coal. On 6 April he wrote Cist that he was not as anxious to be early in the season with his boats as he would have been under other circumstances. He felt there would be many opportunities for running boats during the season. Furthermore, he had not received assurance of getting more than 40 tons of coal. Chapman asked Cist whether he could have coal from the heaps near the bridge. If so, Chapman said he would ship it as soon as the water was high enough. No coal had yet departed from Lausanne. Chapman also wanted a list of people who might be interested in buying coal.

Cist replied the next day, sending Chapman a copy of his statement on Lehigh coal. He had just had 200 copies printed as proof that he remained determined to bring anthracite into use and promote the Lehigh coal trade. Even with the arrival of peace and the opportunity it brought for pursuing other matters, Cist said he saw no business which promised any greater opportunity. He told Chapman he had received orders from two different Philadelphia manufacturers, each for an ark load of anthracite. They offered to pay Cist $15 per ton while at the same time they could get Virginia coal at $7 or $8 a ton. Cist mentioned these facts because he did not want Chapman to feel he had been neglected.[15]

If Chapman made arrangements to run between 140 and 200 tons during 1815, Cist said he could have all the coal on the riverbank. Cist intended to have more coal mined and hauled to the landing. He mentioned that coal merchants along the Lehigh received that spring from $7.50 to $8.20 per ton. In reply to Chapman's question about coal dealers, Cist supplied 12 names at six different places from the Lehigh Water Gap to Philadelphia. The list included a lumber merchant and the keeper of the bridge at Allentown, a distiller and storekeeper at Bethlehem, and a tavern keeper at Easton. In the Delaware Valley, Chapman could leave coal with Joseph Smith, the Tinicum blacksmith, and with three men in New Hope, a lumber merchant, a tavern keeper, and a blacksmith. Cist also gave Chapman the names of his three Philadelphia customers. Chapman should land White and Hazard's coal at a wharf above Vine Street, and 1,000 or 2,000 bushels should go to William Thomas at the South Walnut Street wharves.[16]

During April 1815 Cist received several reports about the demand for anthracite. Joseph Smith reported that he had made little progress keeping Philadelphia blacksmiths burning anthracite after Virginia coal started arriving in large quantities. He became disheartened when he discovered that even one previously strong advocate had abandoned hard coal. Only one of the fires which Cist had fixed remained burning in the city.[17] Blacksmiths in the Quaker City had returned to old habits, giving up the use of Pennsylvania anthracite. In addition to Virginia coal, foreign coal again entered the Quaker port and was used by urban craftsmen.

After recording his efforts to introduce anthracite, the Tinicum blacksmith pointed out that the price of Lehigh coal was too high. A number of blacksmiths told him that the high price discouraged others from attempting to use hard coal. Smith felt that at a lower price Cist would make a greater profit by expanding consumption. He did not think anthracite could be sold in Philadelphia for more than $.625 per bushel. At that time Virginia coal was selling $.05 below the price of Cist's coal.

Joseph Smith gave Cist better news concerning the demand for Lehigh coal in Bucks County. Several blacksmiths in his county had converted to hard coal. To expand its use, he proposed employing a person to work in some of the principal shops in towns and the countryside and paying him a bounty of $1 to $2 for each place in which he introduced anthracite. Smith had a person in mind who understood the combustion of hard coal and was well suited for the job.

In April 1815 Joseph Smith's greatest concern was his inability to get more hard coal for himself. He began to fear that he would not receive much coal that spring. He had not seen or heard of any coal coming down the river. Smith expressed disappointment that two wagon loads which he had acquired came to only 28.5 bushels instead of the anticipated 42.[18]

One month later Joseph Smith wrote to Cist and Miner about the likelihood of securing more Lehigh coal, and he had also asked a relative of Charles Miner to make the same request. Smith said he could easily dispose of a quantity in Bucks County since the demand had increased. Having received no response about receiving a supply, he wanted to know how soon he could acquire a shipment. Because of the efforts Cist and his partners had taken for shipping coal, he had neglected to find a boat or make provisions for getting it by water.[19]

Smith had encouraged one firm to dispose of a considerable quantity of its coal supply in order to introduce anthracite among blacksmiths. This firm would be disappointed if anthracite did not come down the river. Smith's nephew also wanted more hard coal. The previous day he had come to the shop to ask about the prospect of being supplied. His nephew had told the dealers in firewood that he no longer needed charcoal because anthracite produced a superior fire. If he did not get more anthracite, he faced an embarrassing situation in order to carry on his trade.

Smith asked about getting a coal flat that his son had seen in the Lehigh Water Gap. Would Cist send it down to Tinicum as soon as the river rose? If Cist could not meet his request, he wanted an immediate reply so that he could prepare to haul anthracite by land. He also said that if there were any planks for making a flat, he would send his son or come himself to construct the craft.

One thing was certain. Demand for anthracite existed in Bucks County. The Tinicum smith was willing to cart coal by water or land for a distance of about 50 miles to meet the need. Jacob Cist welcomed the news. Three days after receiving his letter, Cist informed Chapman, who was at Lausanne, about the order from Bucks County. Cist told Chapman not to stop at Allentown with the first ark because consumers in the Lehigh Valley could get their coal by wagon much easier than those farther away.[20]

Encouraging news for the hard coal merchants also arrived from Philadelphia. On 17 April Steelwagon and Knight sold two ark loads and two scows of anthracite for $984.34. Cist reported to Chapman that the entire amount in Steelwagon and Knight's coal yard had been sold. Cist was happy to see that the price since peace had stayed at $.75 per bushel, which was higher than he had expected. He said that Philadelphia should not be left without coal for a day. If Chapman did not take coal to the city soon, he would encourage one of his other partners to ship coal downstream to the Quaker City.[21]

Cist let Chapman decide whether to run an ark load of anthracite to Philadelphia, Allentown, or wherever he could get the highest price. If Chapman could get $16–17 per ton with at least half the payment in cash, Cist said it would be better than being detained in Philadelphia with a single ark load where the charges for cartage and commission would have to be considered. Cist concluded his letter by urging Chapman to get flats constructed, oars ready, and the craft either partially or wholly loaded so that he could depart at a moment's notice.

April and May of 1815 were extremely dry months on the upper Lehigh. Chapman waited for the rain to come and the river to rise so that he could send coal downstream, and Cist engaged four men who were willing to take arks down the Lehigh. On 9 May Chapman reported that there was still no prospect of a freshet. Even though the river water remained low, boatmen became anxious to depart. Chapman wrote Cist on 15 May that three flats had left several days previously, but the river was too low. One got stuck in the narrows, and two others stopped below the bridge. It had started to rain hard that evening, and Chapman hoped a freshet would soon follow. With two boats nearly ready to make the trip, he planned to commence loading in the morning. If the water stayed high for three or four days, he could take anthracite to market. There is no indication that Chapman sent coal down the Lehigh River in May.

On 17 June Cist wrote Chapman about a heavy rain and hoped that anthracite arks had gone downstream. As this letter indicates, the greatest obstacles in carrying on the coal business were poor river transportation and insufficient capital. Cist said it cost him more than $1.25 per ton to mine coal. He was willing to sell fine coal at the mine for $2.80 per ton and the large size at $3.00 per ton. Or customers could purchase anthracite for $7.00 a ton at Mauch Chunk or $8.00 per ton at Weiss's house. Cist said he was willing to make a contract to sell anyone any amount of not less than 1,000 bushels annually. The sacrifice Cist had made to establish a market for anthracite had been very great, and he maintained that the enterprise must soon begin to make a profit. Steelwagon and Knight still owed Cist $984.34 for the coal they had sold for cash in April, and he would have to wait until the August court to get paid by another party. Meanwhile, he had paid the wagon masters for hauling coal to the Lehigh.[22]

Jacob Cist and Isaac Chapman dreamed of the day when they could form a company with sufficient capital to carry on the coal trade properly. In his letter of 17 June 1815, Cist told Chapman that his ideas about the business appeared to be very judicious. He promised to consider them carefully. Within two months they drew up articles of association for the Lehigh Coal and Lumber Company. The project would be capitalized at $20,000 with 200 shares worth $100 apiece. Each stockholder would pay $10 per share at the beginning of the subscription, $10 per share on 1 January 1816, and $10 per share every 60 days thereafter until the stock was paid in full. Cist agreed to take 26 shares in exchange for the equipment and effort he had already put into the business as well as his lease with the Lehigh Coal Mine Company. Cist figured the lease was worth $1,600. He would assign the scales, weight, stove grate patterns, and mining tools to the new company. The equipment and his work created a demand for anthracite which he valued at $1,000.[23] Cist gave reasons for forming a large company and assigning his lease to it. He said that only with a greater amount of capital than he had invested could the Lehigh coal trade be carried on profitably. At that time he did not have at his disposal a sufficiently large sum to conduct the business. Besides, his family did not care to move to Lausanne.

To attract capital investment, Cist, as in previous partnerships, set up a balance sheet showing the costs for getting coal to market and the selling price for Lehigh anthracite. He estimated the total cost, involving all expenses, at $9 a ton. In August 1815, Cist's calculations came to $1.50 less per ton than in March 1814, which reflected a drop in postwar prices. He quoted the price of coal in the Lehigh Valley at $15–16 per ton and on the lower Delaware at New Hope, Trenton, and Philadelphia at $16–18 per ton.[24] As shown in the accompanying table, these costs and prices would yield nearly 100 percent profit. An investment of $20,000 in the coal business would permit the company to mine and market at least 2,000 tons of anthracite annually and was sufficient to greatly expand the trade.

Costs and Prices per Ton for the Lehigh Coal Trade

Costs		*Prices*	
Mining	$.75	Selling price at Allentown, Bethlehem, and Easton	$15–$16
Hauling to river	3.00	Selling price at New Hope, Trenton, and Philadelphia	$16–$18
Weighing and loading	.25		
	4.00		
River transportation	5.00		
	$9.00		

Jacob Cist hoped to get backing for the Lehigh Coal and Lumber Company from Connecticut capitalists. Many yankees in Wilkes-Barre had maintained contact with their New England relatives. Money made in mercantile activity during the previous decades was available for reinvestment in profitable projects. Asa Miner, Jr., a New Haven merchant, took an interest in the venture and offered to raise funds for the enterprise. Chapman wrote Cist on 8 August 1815 that he would be happy to see the New Haven merchant and would give him every bit of attention and hospitality which he could afford.

On 19 September 1815 the New Haven merchant sent word to a Wilkes-Barre merchant that he had shown the articles of association to several interested gentlemen and conversed with them about the Lehigh Coal and Lumber Company. But he had met with little success. Their principal objection concerned the short duration of the association. The articles stipulated that the company would last eight years, the remaining years of Cist's lease with the Lehigh Coal Mine Company. Potential investors recognized that the Lehigh Coal Mine Company could choose not to renew the lease and as a consequence put them out of business. They reasoned that the Lehigh Coal and Lumber Company's capital improvements on the roads, navigation, and in every other respect would make the property much more valuable and bring the greatest benefit to the original landowner. If Cist could arrange for a permanent lease or purchase of the property, the New Haven merchant said there would be sufficient encouragement. He had no doubt that many subscribers could be obtained in New Haven who would be willing to enter the business. He expected to be in Wilkes-Barre at the beginning of October to talk about his suggestions.[25] Lack of further correspondence from the New Haven merchant indicated that Cist failed to secure capital from this city. Once again his plans for a heavily capitalized corporation to carry on the Lehigh coal trade failed to materialize. He would have to look other places for investment funds.

Without sufficient capital to carry on a large-scale enterprise, Cist and his partners resorted to conducting the coal trade on a part-time basis and on a small scale. They relied on Isaac Chapman to live at Lausanne and look after the business. Chapman, who was engaged in two or three other projects, stayed at Lausanne when not pursuing his other work. Cist kept Chapman supplied with provisions from Wilkes-Barre and only occasionally journeyed to the Lehigh coal mine. Prices for anthracite, however, remained high enough to let them stay in business.

On 15 November 1815 Isaac Chapman wrote Cist about plans for the coming winter. Chapman was willing to employ two men at $20 per month provided they would chop logs, dig coal, and work at other projects. If Cist hired the two men Chapman had in mind, the one could be used to pilot arks down the Lehigh and the other would make an excellent boatman. Pay for their work while rafting would be higher than their normal wages.

Chapman planned to go ahead with capital improvements. He had started opening a mine at Nesquehoning and building a road from it to Lausanne. Two coal pits had been cleared which Chapman felt were part of the same vein.

In November Chapman expressed fear that there would not be any freshet before spring to take coal to market and deplored the fact that nothing further had been heard from New Haven. Chapman's fear came true. He wrote Cist on 25 December about a small freshet on the Lehigh which brought the river about halfway up the rafting rock. He said that this would be the last freshet of the year. To date, no coal had left the landing, and Chapman gave up hope of running any coal flats that season.[26]

Calling himself the Jonah of the enterprise because bad luck had plagued the partners ever since he had joined them, Chapman gave Cist another piece of bad news. One party, upon learning that there was no chance of running his coal arks that season, refused to pay Chapman on his contract. Chapman would be forced to take the matter to court in order to get his money. Chapman also told Cist that someone was burning Charles Miner's coal. And Nathan Beach, the operator of a mine at Beaver Meadows, was selling his coal to teams at $7 a ton, or $3 less than Cist's price per ton at Mauch Chunk.[27]

In spite of these difficulties, Chapman went ahead with plans for the spring season. He had made progress in opening the Nesquehoning mine and contracted for the digging of 100 tons. He had arks ready to ship coal as soon as the river permitted boating. Meanwhile, Jacob Cist had kept account of the urban fuel demand. As noted previously, on 22 December 1815 James Ronaldson wrote to Cist from Philadelphia that fuel prices continued to rise. He believed that the city's residents spent more than $1 million a year for fuel and said that urban craftsmen paid nearly two-thirds of this amount, making the total fuel bill nearly $2 million a year. Ronaldson said that demand existed in Philadelphia for investment in transportation and mining companies that would introduce anthracite.[28]

Such news encouraged Cist and Chapman to continue in business. On 4 February Chapman wrote his partners that he was sending specimens of hard coal from the Nesquehoning mine. At last he had found coal of a good quality even though he had to dig much deeper than expected. Now that 60 feet of strata had been removed, mining could begin. After 100 tons had been dug, Chapman felt he would be better able to calculate for the future. Chapman told Cist to get exclusive rights to this mine.

Since the mine was only four miles from a coal landing, the partners could conduct business at a lower cost and sell anthracite cheaper. They were already facing competition in the business from other individuals. Jacob Weiss had sold three tons of coal at $7 per ton. Francis Weiss told Chapman that teams came up for Beaver Meadows coal, which was also selling at $7 per ton. Francis Weiss wanted to know whether Cist's coal could be sold for less than $10 a ton.

The spring of 1816 brought better luck to Chapman and Cist. Wet weather in May permitted coal arks to descend the Lehigh. On 28 May two coal arks arrived at New Hope where Joseph Smith's nephew immediately purchased the entire cargo of 12.75 tons. He paid cash for 10 tons at $18.50 per ton and took the rest on 90-day credit. Chapman received $234.39 for the coal. During September 1817, a smaller shipment on two coal flats went to two Bethlehem merchants. During 1816 and 1817 the Lehigh coal business amounted to less than $500 each year.[29] Although Cist and Chapman managed to mine some coal each year, they conducted the coal trade on a part-time basis.

More profitable employment in 1816 attracted Chapman to move from Lausanne to Wilkes-Barre. At the beginning of the year he purchased Charles Miner's *Gleaner* and went into the newspaper business. Chapman also opened a land office in the Luzerne County seat and received a commission from Pennsylvania to make the earliest maps of Luzerne and Northampton counties. These activities occupied most of his time and left little free time to carry on the Lehigh coal trade. In August 1817 Charles Miner moved to West Chester to run the *Chester and Delaware Federalist* and engage in politics. As a result, he was unable to replace Chapman on the Lehigh.[30] Jacob Cist's preoccupation with Matthias Hollenback's business in Wilkes-Barre and the refusal of Jacob's family to move to Lausanne gave someone else the opportunity to manage the Lehigh coal trade.

Nevertheless, the efforts of Cist, Chapman, and Miner to establish a market for Lehigh coal had not been completely wasted. By investing time and money in capital improvements and by keeping the mines, roads, ark yard, and buildings in fairly good condition, they laid the groundwork for their successors. At least two Philadelphia coal merchants were willing to handle Lehigh anthracite, and a number of urban craftsmen would purchase their product when a sufficient amount arrived in the city at a reasonable price.

The Commonwealth of Pennsylvania provided additional incentive for someone to take charge of the Lehigh coal trade. However, the biggest obstacle in the way of opening the coal trade was poor river transportation. Low water in the river had frustrated Chapman's attempt to market anthracite, and the Pennsylvania legislature was aware of this problem. In 1814 the state legislature passed a law to incorporate a company for improving river navigation, and Jacob Cist was named a commissioner to collect subscription money in Luzerne County. Two years later the legislature passed a supplementary act to aid the company. The biggest boost to this movement came in March 1817, when another law authorized the governor of Pennsylvania to subscribe 500 shares of stock in this company for the Commonwealth. And $50,000 would go a long way toward getting the project started.[31] Money had been appropriated to attract private investment into the enterprise.

The opportunity to engage in the coal business, the availability of state funds to attract capital investment, and the desire to acquire Lehigh anthracite provided sufficient inducement to the entrepreneurs who ultimately created the heavily capitalized Lehigh Coal and Navigation Company and thereby fulfilled Cist's plans. On 7 January 1818 Jacob Weiss, Jr., informed Cist that Josiah White, Erskine Hazard, and George Hauto had come to the coal mine and Lehigh River. Weiss reported that they were very pleased with the capital improvements made by Cist and his partners and felt that river navigation would not be an obstacle. Weiss felt they would take a lively interest in improving the river. Josiah White told Thomas Weiss that Lehigh anthracite far exceeded the quality of Schuylkill coal and felt that Lehigh coal could be taken down river at a cheaper price than anthracite from Schuylkill County. Evidently, White considered the improvement of the Schuylkill River impractical because more capital was needed than the Schuylkill Navigation Company had the power to command in order to complete the project. Due to Cist's connection with the Lehigh coal trade, Weiss felt he should have this news. Weiss also suggested that it might be worth Cist's time to correspond with White.[32]

Josiah White was certainly qualified to undertake improving navigation on the Lehigh River. His background is well known to students of the anthracite industry and need be repeated only briefly.[33] Born in

New Jersey in 1781 of Quaker parents who were connected by kinship with several prominent Philadelphia Quaker families, White, at age 15, became the apprentice of a Philadelphia hardware merchant. As a clerk in the store, he learned to keep books, look after the merchandise, and take inventory. After serving five years, White asked that his apprenticeship be terminated so that he could buy out another merchant and engage in his own business. White wanted to make a profit of $40,000 by age 30, thus becoming independently wealthy. He started with an estate of about $6,000 and within seven years had attained his goal.

While Josiah White hoped to invest half his fortune in a profitable enterprise and with the other half become a gentleman farmer, an opportunity became available in 1810 which caught his imagination. A Philadelphia gentleman offered for sale the Falls of the Schuylkill, the closest water power site to the city, along with three or four acres on the east side of the river and seven or eight acres on the west side. At this point the river fell three and one-half feet and provided ample power to run several waterwheels and a factory. White paid $14,000 for the property as well as the right to construct a lock and charge each boat $.50 to go through the lock.

The Schuylkill River, one of the busiest streams in Pennsylvania, was among the main routes for bringing agricultural produce from fertile southeastern Pennsylvania counties to Philadelphia. No stream as big as the Schuylkill had ever been dammed in Pennsylvania. From 1810 to 1811 White spent a major part of his fortune damming the river and erecting two locks.[34] While boats passed through his locks and earned some revenue, this project had a reverse effect. Farmers and upriver merchants disliked paying $.50 a boat to take their produce through these locks to the Quaker City.

In 1812 Josiah White petitioned the Pennsylvania legislature to incorporate the Schuylkill Navigation Company and make him the head of the corporation. Immediately, Berks, Chester, Montgomery, and Schuylkill counties protested White's proposal. Although citizens in these counties believed in the value of the project, they did not want it to be controlled by Josiah White. After three years of political bickering, the legislature finally incorporated the Schuylkill Navigation Company.[35] White and Hazard subscribed to 20 shares of the company and thereby committed $1,000 of their capital toward the project. This was more than twice the amount subscribed by most Philadelphia merchants who took three, four, or five shares at $50 a share. By 1815 White and Hazard wanted to see the navigation completed so that anthracite could be brought cheaply from Schuylkill County to their factory.

From the outset Philadelphia merchants controlled the Schuylkill

Navigation Company. Neither White nor Hazard was elected to the board of managers, but one of their friends served on the board for the first two years. The original act called for a capital investment of $500,000 to improve navigation from the head of the river to Philadelphia.[36] Being short of capital, the managers asked interested businessmen to construct dams, locks, and canals at specific sites for the company in return for water power rights. Josiah White was the only one to accept the offer. On 14 August 1816 he signed an agreement to construct a dam high enough to make a three-foot pool of water stretching from tidewater four miles upstream and to build locks and a canal along the west bank. In return White would retain all the water power from the falls. He probably entered into this agreement in order to protect his original investment at the falls. White's work failed to completely satisfy the managers, and he acquired a bitter attitude toward the company, which lasted his lifetime. His dam and canal project was purchased by Philadelphia in 1819 when the city agreed to construct Fairmount Dam.[37] By this date White and Hazard had become disenchanted with the Schuylkill Navigation Company and refused to invest any more money in a project which upon completion in 1825 cost over three and a half times the original estimate.[38]

Like Jacob Cist, Josiah White acquired a firm belief about the potential impact of anthracite on American economic growth. He tried to make an agreement with the Schuylkill Navigation Company concerning the toll for conveying anthracite to Philadelphia, but the company refused to accept White's offer. When this proposal failed, White began giving serious consideration to the Lehigh coal trade. Thanks to his experience in constructing dams, locks, and canals, he felt that he could succeed in improving navigation on the Lehigh River. And he was willing to move to the Lehigh region in order to assume management of the enterprise. Upon arriving at the Lehigh, White devised a clever method of piling stones in the stream to make artificial freshets.[39]

No doubt, Jacob Cist was pleased to learn that Josiah White wanted to undertake improving navigation on the Lehigh River and promote the anthracite trade. Finally, the plan which he and Isaac Chapman had formulated in 1815 to create the Lehigh Coal and Lumber Company might be realized. When he had dealt with White and Hazard during the fuel crisis, they had gained his respect. Since he did not want to move to Lausanne to assume management of the company, Cist was willing to let someone he trusted undertake the responsibility. As the principal stockholder in the Lehigh Coal Mine Company, he was willing to grant his colleagues their request and give them the help they needed.

On 29 January 1818 Josiah White and his two partners in the new enterprise signed a contract with the Lehigh Coal Mine Company. They agreed to pay all taxes and other expenses on the property as well as give the company one ear of Indian corn annually for a period of 20 years in exchange for the right to mine and take the coal to Philadelphia. In March they secured an act from the state legislature to improve navigation of the Lehigh and thus began their efforts to assume responsibility for the Lehigh coal trade.[40]

In April, White and Hazard left Philadelphia to survey the river and mine road. They did not have much money to do the work and even had to borrow a surveyor's level. On 27 April Isaac Chapman left home early and arrived at Lausanne in mid-afternoon. Learning that White and Hazard were at Lehighton, he left Lausanne immediately and found the partners. Chapman recorded in his journal that he spent the entire evening talking with them about the Lehigh coal trade. Since he had been involved in this business for five years, Chapman possessed valuable information that he was willing to share with White and Hazard.[41]

To achieve their goal, White, Hazard, and Hauto realized they needed to establish a heavily capitalized corporation. The three partners proposed to create the Lehigh Navigation and Coal Mine Company which would be capitalized at $200,000 with 200 shares of stock, each worth $1,000. In exchange for the mine lease and river rights, they would each retain 50 shares and control of the company. They planned to sell the remaining 50 shares to raise capital for the improvements and stated that these shares would receive the first payment of dividends.[42]

Acquiring the necessary capital proved to be a difficult task. White had kept his project secret until funds were needed. The arrangements called for George Hauto, who was reputed to be a German mining expert with a considerable fortune, to raise the money. Much to his surprise, White found that Hauto was a fraud with a poor credit rating among Philadelphia financiers. Consequently, the task of selling the stock fell on White and Hazard. According to White's journal, the partners visited 19 of the city's leading investors, and all rejected their offer. White wrote to Jacob Cist in Wilkes-Barre, hoping he would subscribe to some of the stock, but a lack of capital in the frontier town prevented the Wilkes-Barre merchant from fulfilling the request.[43]

Although Cist did not invest in White's project, he may have proved instrumental in helping him get the backing of John Stoddart, one of Philadelphia's leading merchants. Stoddart had established a substantial trade with the German population of Northampton County, owned a large tract of land along the upper Lehigh, and in 1817 had offered to subscribe 25 or 50 shares in a bank which Matthias Hollenback wanted to set up in Wilkes-Barre. Cist and Hollenback dealt with Stoddart on a number of occasions, sending large shipments of country produce down the Lehigh Valley to Easton and Philadelphia. Cist most likely prevailed upon John Stoddart to help finance White and Hazard's project. By 1818 Cist had finished his scientific investigation of the anthracite formation and made plans to connect the northern coal formation with the Lehigh River. His work showed that a large volume of anthracite would descend the Lehigh once a permanent means of navigation had been created. No doubt, Cist's efforts helped persuade Stoddart to back White and Hazard.

Josiah White's memoirs give credit to John Stoddart for influencing other Philadelphia financiers to invest in his projects. After failing to get support for the Lehigh Navigation and Coal Company, White and Hazard proposed setting up two separate companies to achieve their goal, the Lehigh Navigation Company and the Lehigh Coal Company. In August 1818 Stoddart subscribed $7,000 to the Lehigh Navigation Company, and 19 other investors took the remaining shares. Two months later Stoddart subscribed $5,000 to the Lehigh Coal Company, and seven others raised the rest of the necessary capital.[44] An initial start had been made, even though more money would be needed. After two more reorganizations the Lehigh Coal and Navigation Company was finally chartered in 1822.

After eliminating Hauto as a partner, White and Hazard moved to Mauch Chunk and undertook the responsibility of making the Lehigh River a slack water navigation system. Indeed, Josiah White's greatest achievement was constructing the 46-mile slack water navigation system from Mauch Chunk to Easton. In addition, White and Hazard constructed the other facilities planned by Cist and Chapman to provide Philadelphia with a steady supply of anthracite. In 1820 they sent 365 tons of hard coal and four years later shipped nearly 10,000 tons of anthracite to market in Philadelphia. According to White, the turning point had been reached because the Quaker City had become dependent upon Pennsylvania anthracite for its fuel.[45]

In the 1820s White and Hazard created one of the largest private corporations in the United States, and they became the general managers. As the size of their business grew, they needed experts who possessed knowledge concerning special aspects of their enterprise. In 1826 they acquired the services of Isaac Chapman to supervise their mines. Chapman's knowledge of the geological formation and his experience as Cist's agent qualified him for the position of mining superintendent. Unfortunately, Chapman became ill the following year and died in 1827 at age 40.

Jacob Cist never lost interest in the Lehigh coal trade. His cousin Francis Weiss continued to engage in the Lehigh trade after Cist and Chapman had left the scene in 1818. In October 1820 Francis Weiss wrote to Cist that his brother, Thomas, received little cash for his coal and had to sell it as best he could because Josiah White sold anthracite at very low prices in Mauch Chunk. Coal prices had fallen from $10 a ton at Mauch Chunk to $3 and $4 per ton for the best quality.[46]

In Philadelphia the Lehigh Coal and Navigation Company sold anthracite for $.30 a bushel or $8.40 per ton. Liverpool coal, on the other hand, sold for $.43 a bushel in New York City, or $.13 a bushel more than Lehigh coal. Many Philadelphians converted to Lehigh coal, and as previously noted in Cist's pamphlet, they considered it superior to both Liverpool and Virginia coal as well as firewood. For the first three years of business, White had some hard coal left over in the spring. But by 1824, when nearly 10,000 tons went to market, most Philadelphia consumers realized that an ample supply would come to market regularly.

In the meantime White and Hazard had created conditions for a potential rival to enter the business. They had set up their own coal wharf on the Delaware River and proceeded to distribute anthracite from door to door throughout the city at $.30 per bushel. In reporting this practice to Cist in December 1821, Frederick Steelwagon said that the Lehigh company had alienated the old coal dealers with whom he had worked during the fuel crisis. White and Hazard limited their sales to housekeepers and excluded the coal yards until all others using hard coal were supplied. Steelwagon told Cist that if Pennsylvania anthracite could be supplied to his coal yard at any level near White's price and not in excess of $.35 per bushel, he would give the Lehigh Coal and Navigation Company competition. Cist indicated a desire to supply Steelwagon and Knight once again with anthracite.[47] Not only had Cist helped develop the Lehigh coal trade, he was also willing to create competition against it.

Historical accounts give Josiah White and Erskine Hazard sole credit for establishing the Lehigh coal trade, overlooking the preliminary work done by Cist and Chapman. White and Hazard deserve recognition for the permanent navigation improvements made on the Lehigh River and for creating the Lehigh Coal and Navigation Company. But their predecessors made important contributions necessary for their success. Besides getting merchants to handle their product and establishing a market for anthracite among Philadelphia craftsmen, Cist and Chapman had made capital improvements on the mines, roads, and coal landing as a basis for White and Hazard to start their work.

5 Evolution of the Wyoming Valley Coal Trade

Even before relinquishing his rights in the Lehigh coal mine to White and Hazard, Jacob Cist had become deeply immersed in the development of the Wyoming Valley coal trade. The policies followed by the firm of Hollenback and Cist add another dimension toward understanding the process of establishing a market for Pennsylvania anthracite. Initially these policies had been created to enlarge Hollenback's business in an agrarian economy, but they were easily adaptable to the anthracite trade.

In 1834 George M. Hollenback jotted down instructions for a new assistant to help manage his business in Wilkes-Barre. The firm's memorandum and account books and other fragmentary manuscript material indicate that these rules had been followed by Cist and Hollenback since the War of 1812. Except for the points concerning coal, their enterprise did not differ too much from other American merchants during this decade.[1]

Hollenback listed eleven points for his assistant in order of priority.[2] Liberty has been taken to paraphrase them in order to clarify their meaning.

1. Collect all old debts which can be acquired.
2. Settle all open accounts.
3. Attend carefully to the store.
4. Keep the gristmill running with whatever there is.
5. Keep the sawmill and another mill in good condition and operation.
6. Whenever the team can do nothing better, have coal hauled to town and to the riverbank.
7. Clear the wood from a new piece of land, and sell the wood.
8. Cut the trees on the new road to the coal mine. Prevent anyone from taking the trees.

9. Make sure there is plenty of coal on piles at the mines so that people coming to the mines will have a ready supply.

10. Haul boards to the sawmill to make shingles.

11. (illegible).

As can be seen, Hollenback's motive was to make money. Collecting old debts and settling accounts ranked first in importance. Bad debts were money lost and could lead to insolvency. Hollenback's reputation as a businessman rested, in part, on his ability to collect money, and he was successful. Matthias's memorandum books from 1813 to 1825 prove this point.[3]

Points 3 to 5 show the main way in which Cist and Hollenback made money. As merchants in an agricultural and commercial region, they engaged primarily in exchanging country produce such as grain, cattle, and timber for manufactured goods in Philadelphia, Baltimore, and New York City. Their stores on the upper Susquehanna River became centers of community activity. The extensive forests and rich farm land of the Susquehanna Valley yielded products which were marketed by the Hollenback firm. They reinvested excess capital in several mills to process the products of farm and forest. Most of the transactions in their record books deal with agricultural goods and merchandise brought from Philadelphia. Their books also show the initial transition from an agrarian economy to one in which anthracite would become the leading sector.

The Wyoming Valley had certain advantages which helped make it competitive with the two major centers of hard coal production south of Broad Mountain prior to 1820. This valley was the only rich agricultural area within the anthracite formation, and Wilkes-Barre was the only town until Mauch Chunk and Pottsville became urban centers in the 1820s. The population of the Wyoming Valley had grown steadily since the American Revolution. According to the federal census, in 1820 Wilkes-Barre and the six townships making up the valley had a total population of over 7,000, three times more than the townships producing coal in the southern field.[4]

After the invention of Jesse Fell's grate, the farmers in the Wyoming Valley used hard coal to cook and heat their homes. If they did not have an anthracite outcrop on their farms where they could dig their own fuel, they could barter goods and services with at least six mercantile firms in the Wyoming Valley. In an area scarce of capital, their business helped sustain the growth of the Wyoming Valley coal trade.

To acquire a comprehensive picture of the coal business in the Wyoming Valley requires going beyond the account books of Hollenback and Cist. Many merchants and farmers opened outcrops on their land and engaged in the coal trade. Charles Miner's *Gleaner* posted notices of several coal mines for sale during Philadelphia's fuel crisis. After the war similar advertisements appeared. For example, during May and June 1815 an advertisement in the *Gleaner* offered to sell the right to raise and take anthracite from property along the riverbank in the lower part of Pittston and also from the land of two proprietors in an adjacent township. Similar notices continued to appear throughout the period.

In dividing up the Wyoming Valley, the Susquehanna Company provided for numerous small farms. Anthracite outcropped on many of these farms. Several farmers in the valley opened coal beds in their spare time and used this fuel to heat their own homes and sold small amounts to neighbors. These mines were the smallest units of production. According to an 1830 map of the Wyoming and Lackawanna valleys made by the operator of the famous Baltimore Coal Mine, there were nearly 100 coal mines or outcrops. He put the names of the owners by 36 of the mines, just about every one of which was known to produce coal by this date.[5]

Even if every one of these mines produced anthracite during these years, there were only about six firms that sold coal on any extensive scale. The largest operation belonged to Abijah and John Smith of Plymouth, who began their business in 1808. Throughout this period, Jacob Cist ran the coal business of Matthias Hollenback, and his chief Wilkes-Barre competitor engaged in the business. On the north side of the river two Plymouth merchants and two Kingston merchants had also gone into the trade as had a Nanticoke ironmaster. Sufficient records have survived for two of these concerns, the Smith brothers' and Hollenback's, to show the growth of the local trade. Fortunately, these records also provide substantial information about the coal business of their competitors.

Capital to finance the coal trade prior to the 1820s was extremely scarce. As a matter of fact, it was the scarcest item for anyone to acquire in the early national economy. As discussed previously, Cist and Chapman were unable to get sufficient capital for their Lehigh enterprise, and Josiah White also had a difficult time getting financial backing for the Lehigh Coal and Navigation Company.

In the Wyoming Valley most farmers found it relatively easy to mine and market a few tons of anthracite a year since they already owned the necessary equipment: a pick, shovel, team, and wagon. Clearing debris off the face of an outcrop, using a pick and shovel to loosen chunks of coal, and putting them into a wagon were simple tasks. Limited financial resources, however, prevented farmers from doing any extensive local business. Only merchants in the villages of Kingston, Nanticoke, Pittston, Plymouth, and Wilkes-Barre had sufficient capital to pioneer the trade, and each town had at least one or two merchants willing to invest their excess profits.

The case of Hollenback and Cist was typical. They made money from various activities, collected on their bills, and kept the store and mills open and operating at all times. The anthracite trade offered another profitable opportunity for expanding business, and they invested excess capital in improving the mines and transportation facilities as time and money permitted. A similar situation can be seen in the accounts of John Smith, who traded regularly with the stores of two Plymouth merchants and two other merchants in Kingston.[6] The nature of their business resembled that of Hollenback and Cist in Wilkes-Barre. Through the gradual growth of business, these merchants planned to accumulate capital to develop the anthracite trade.

The Susquehanna River offered another major advantage to Wyoming Valley merchants competing with anthracite produced in the Lehigh and Schuylkill valleys. Even though hard coal had to travel a longer distance down the Susquehanna to the Philadelphia market, the river was larger and did not have as sharp a descent compared with the Lehigh and Schuylkill rivers. Consequently, navigation on the Susquehanna was much easier.

The price structure for shipping anthracite down the Susquehanna River during the decade after the War of 1812 provided sufficient incentive for merchants to enter the coal trade. The accompanying table shows the various costs and profit. Although $145–160 profit on one ark carrying 60 tons does not appear high in light of present-day prices, to businessmen of that period it appeared good when compared with wages or the cost of land. Common labor in the Wyoming Valley earned only $.50 a day, or $180 annually. Skilled labor cost twice as much as common labor. In 1815 good Luzerne County farmland sold for $3–6 an acre, or after 1820 a person could buy an 80-acre farm from the federal government in the Midwest for $1.25 an acre, or $100.[7] Of course, the cost of shipping anthracite from Havre de Grace to Philadelphia or another Atlantic coast port reduced the profit still further, but Wyoming Valley merchants found the profits sufficiently high for them to engage in the coal trade and send one or two arks of anthracite down the Susquehanna River annually.

The policies of George M. Hollenback, mentioned at the beginning of this chapter, became the rules for all businessmen wishing to engage in the Wyoming Valley coal trade. To survive financially, merchants and other individuals became mutually interdependent, and each of Hollenback's points helped them develop the anthracite trade.

Costs and Profit of Shipping Anthracite Down the Susquehanna[8]

60 tons of coal @ $1 to $2 per ton on the riverbank	$60–120
Coal ark—90 feet long by 16 feet wide and 4 feet deep	50
Susquehanna River pilot ($60 to $75)	60–75
3 raftsmen for an 8-day trip plus expenses	30
Cost of getting 60 tons to Columbia or Port Deposit	200–275
Price of anthracite at Columbia or Port Deposit, $6–7 per ton × 60 tons	360–420
Profit	160–145

To engage in the coal trade one had to secure credit from three months to one year in order to market the fuel and at the same time meet other obligations. At least one local resident wishing to enter the coal business turned to the Hollenback firm to secure sufficient capital. On 18 January 1823 he made an agreement to borrow from $600 to $1,000 in goods, thereby enabling him to ship anthracite down the Susquehanna River. If he repaid the loan promptly, he would receive a discount of 5 percent on the goods. For security, he put up the deed for his property. In this way he managed to become a coal merchant.

If merchants failed to collect on current debts, they often had to ask for an extension of credit. For example, on 23 July 1821 Abijah Smith wrote to George Hollenback about postponing payment on two of his notes. He told Hollenback that during the previous week he had gone to the west bank of the Susquehanna to collect $647 owed to him. He could not collect any money there but received a promise he would be paid in the fall. Smith said that it was out of his power to settle any sooner and asked Hollenback to wait for the money.

This constant cycle of indebtedness as well as the ability to collect debts were important elements in the economy. It is no wonder that Hollenback placed this point at the top of his list. Anyone who failed to heed it soon went out of business. Success in the coal trade required that merchants follow this rule religiously.

On several occasions Wyoming Valley merchants helped each other sell coal and collect fuel bills, particularly when they were in a position to do so. Luzerne County elected George M. Hollenback to the 1824–1825 session of the Pennsylvania legislature. While in Harrisburg, Hollenback received a letter from John L. Butler, his leading competitor in Wilkes-Barre, that an ark carrying more than 60 tons of anthracite had become stuck on an island in the river and that a nearby tavern keeper had offered to purchase it. He had dealt with the tavern keeper previously and did not trust him. If Hollenback could get $160 for the 60 tons as it lay on the riverbank, he was willing to make the sale. He would wait until spring for the money. Otherwise, the Wilkes-Barre merchant told Hollenback to advertise the coal in a Harrisburg newspaper and sell it that way if he could.

Two months later George Hollenback received a letter from Henderson Gaylord, a Plymouth merchant, who thanked Hollenback for attending to the sale of a 56-ton ark load of anthracite. He asked the legislator to call upon the purchaser once again for a note or money to pay for the coal. He felt that his reputation was at stake if the bill went unpaid. For this service, the Plymouth merchant said he would be deeply indebted to Hollenback.[9] Providing such services was one way that local politicians built up their base of political support and broadened their influence. At the same time they kept afloat the complex system of debits and credits in which they were involved. Sooner or later they might have to call on a fellow merchant to render similar service.

Running their local stores became the major means for these merchants to make money. Anthracite was only one of numerous items sold, for just about every merchant in the valley handled hard coal.

Anthracite was not the only item sold at Hollenback and Cist's store which helped stimulate the trade. The Hollenbacks took a special interest in renting and selling coal stoves. In October 1817 Cist's old friend Judge John B. Gibson wrote to George Hollenback about the possibility of selling a coal stove which he had previously purchased from the merchant. He said that it was well constructed for burning anthracite and should bring a reasonable price since it was autumn. He asked Hollenback to take it to the store and put it in view of the public. Since it had not been used very much, he wanted to sell the stove at the original price, but he was willing to accept a lesser amount.[10] This stove appears to be one of the earliest sold in Wilkes-Barre. Most homes still depended on fireplaces for cooking and heating and used Jesse Fell's grate to burn anthracite.

Conversion to stoves in the Wyoming Valley came during this decade. On 16 November 1818 Reuben Trelxer, an ironmaster, wrote to Matthias Hollenback from Mary Ann Furnace near Bethlehem that he had all sorts of wood and coal stoves on hand. If Hollenback wanted any stoves, he would be happy to supply them. The price was $95 a ton for stoves, and Hollenback could buy them on nine- or twelve-months' credit. He was also willing to accept boards and roofing shingles in exchange for the ironware. In December 1818 the Wilkes-Barre merchant purchased several wagon loads of coal stoves.

During the winter of 1818–1819, the Hollenback firm sold these coal stoves at their Wilkes-Barre store. They cost the company about $10 apiece, and Hollenback and Cist sold them at roughly twice that price. Between 1818 and 1830 the Hollenbacks became a major supplier of coal stoves in northeastern Pennsylvania. They acquired most of their stoves from the Bethlehem ironmaster whose two most popular patterns had been designed by Jacob Cist (see figures 6 and 7). But the Bethlehem ironmaster was not the only one interested in furnishing Hollenback's store with coal stoves. In October 1826 George Hollenback received a quotation from another Lehigh Valley furnace for coal stoves which was $10 per ton higher than the price given by the Bethlehem ironmaster.[11] Consequently, Hollenback placed his order with the Bethlehem forge.

As early as December 1818, Jacob Cist wanted to get the molds for his coal stoves so that they could be produced closer to home. Cist owned the rights to several iron ore deposits in the Susquehanna Valley and knew most of the ironmasters. If they could make his stoves as cheaply as the Bethlehem ironmaster, Hollenback and Cist could save the cost of wagon transportation from Mary Ann Furnace to Wilkes-Barre. Cist felt that the Bethlehem ironmaster owed him two coal stoves so he could make the patterns.

On 2 December 1818 the Bethlehem ironmaster wrote Matthias Hollenback that he did not remember ever promising Cist the two coal stoves. If Cist could convince him that they had made such an agreement, he was willing to settle the debt. Cist ordered stoves to be cast according to his pattern which the ironmaster said would be made in the spring of 1819. A disagreement arose as to who would pay for them. In September 1821 the ironmaster again informed Hollenback that Cist had failed to prove that he owed the inventor these stoves. He said that Matthias should collect the money for the stoves from his son-in-law.[12]

George Hollenback's memorandum books show that Cist's stoves were sold in the Wyoming Valley, and that local craftsmen helped produce these stoves. On 22 December 1826 Hollenback sent ten coal stoves to one Wilkes-Barre shop to be finished at $1 each. The craftsman had to put iron feet on the stoves and fasten sheet iron to the sides with wire. If their books are an accurate indication, the Hollenback firm increased their sale of coal stoves from between five and ten a year in the early 1820s to nearly 30 in 1830.

Renting coal stoves for the winter offered merchants another means for making a profit and increasing coal sales. Most people in the Wyoming Valley were too poor to buy stoves, but often they could barter goods and services for their winter's fuel supply and a stove to keep warm. On 9 December 1819 Hollenback and Cist let one local resident have a ten-plate stove, one elbow, and five pieces of stovepipe for the winter. He was to work for the company until he earned $5, the rental price for the stove. At $.50 a day he would repay the debt in ten days. The next year they rented the stove to another Wilkes-Barre resident. George M. Hollenback continued the rental policy, having from five to ten stoves leased each year from 1825 to 1830.[13]

By 1830 a large number of people living in Wilkes-Barre kept their homes warm with one of Cist's coal stoves and purchased their fuel from George Hollenback. Other people in the Wyoming Valley heated their homes with anthracite stoves supplied by other coal merchants. George M. Hollenback also sent coal stoves to nearby towns. In October 1828 a merchant in Berwick asked George Hollenback to send him a small stone coal stove to heat his store since he could not get one in Pottsville. The next month he asked Hollenback to send a larger coal stove for the canal office. In 1830 Hollenback received another order for a coal stove from 30 miles up the Susquehanna River at Montrose.[14] Anthracite was being burned to keep houses warm up and down the Susquehanna Valley. Jacob Cist's prediction that sooner or later nearly everyone in eastern Pennsylvania would use anthracite-burning stoves was rapidly coming true.

Bartering goods and services at stores in the Wyoming Valley became the major way of paying for anthracite and stoves and made up a sizable part of the merchants' trade. From 1815 to 1825 John Smith recorded 258 entries in his books for the sale of coal in the Wyoming Valley. Of these, 126 entries were for one or two loads with each load amounting to 1.25 tons. Fireplaces consumed far larger amounts of fuel than stoves which is reflected in the fact that John Smith also had 15 steady customers who took four or five tons of anthracite annually. In addition, he dealt with four blacksmiths who purchased an average of six tons a year.[15] While Smith sent some coal in wagons to Wilkes-Barre, most of his sales were on the north side of the river opposite the town. Hollenback and Cist and other Wilkes-Barre stores peddled most of their coal on the south side of the Susquehanna.

Most customers paid their fuel bills by working a few days or weeks for the coal dealer. As part of their wages, Hollenback and Cist gave full-time employees the privilege of getting their own coal at the mine. For example, in December 1815 one employee took the yoke of oxen to Matthias Hollenback's coal bed to get his winter's fuel supply. The account books contain similar entries for three other full-time employees who were general handymen. To get a Northampton County cooper and a miller to come to work for him, George M. Hollenback made agreements that they both had the right of getting coal for family use.[16] Since fuel bills were a major part in the cost of living, these privileges were attractive and helped draw scarce labor to the frontier community.

Others worked shorter periods of time for their fuel. During slack winter months farm labor had excess time which could be used for mining and hauling coal. Most of the anthracite was mined during the winter months of December, January, and February. Coal mining paid higher wages than common labor because it was considered one of the hardest jobs. Jacob Cist and George Hollenback paid most of their miners for the number of tons dug or on shares instead of a daily basis. In 1818 one miner received $.75 per ton for 132 tons. When mining on shares, workers usually received one-third of their total output. One man could dig about seven tons a day. Under such a system he earned $3.50–4.00 per day. In this way Cist and Hollenback obtained at least 100 or 200 tons of coal annually from 1815 to 1830.[17] John Smith used the same methods even though his annual output far exceeded the Hollenback firm.

Most farmers had teams and wagons or sleds which could haul coal for the merchants during the winter months. From October until April of each year, a steady stream of wagons traveled the road from the Mill Creek mine to the riverbank and Wilkes-Barre or carted coal from Smith's mine to nearby villages and farms. John Smith established regular delivery rates at about $.01 per bushel per mile. He had coal carted anywhere in Plymouth for $.02 per bushel. The charge to Kingston was $.03 per bushel, to Wilkes-Barre $.04, and to Berwick $.05 or $.06. Of 34 regular coal customers Smith had working for him, nine were employed to haul coal. At $1 a day wages, a person with a wagon could pay his annual fuel bill with one week's work. In addition, a number of farm wagons brought produce to the country merchants and carried away tons of anthracite.

Craftsmen contributed valuable services in exchange for their fuel. All four blacksmiths dealing with John Smith received coal in return for their work. Between 1815 and 1818 two blacksmiths took a total of 27.5 tons of anthracite. To pay the bill, one of them fixed two wedges, sharpened two picks, and welded one pick for use in the mine. He also straightened one pair of bent chains, shod the horses, and fixed a grate for burning coal, while the other blacksmith fixed a plow and did other work. The job of fixing and sharpening tools as well as shoeing horses was a continuous task as indicated in the accounts of the other two blacksmiths who worked for John Smith from 1818 to 1824. For example, one of the men needed 30 tons of anthracite during these six years, and he paid his coal bill by working in the blacksmith shop.[18]

Carpenters also found ready employment in John Smith's ark yard. Skillful carpenters commanded $1 a day wages to construct the 90-foot-long by 16-foot-wide and 4-foot-deep craft. From 1815 to 1825 John Smith employed a total of six carpenters to build his coal arks. Four of them worked during the summer months, indicating that they followed the carpenter's trade full time; two others were more likely farmhands who turned to building boats during the winter. Each of the carpenters took coal as part of his pay, with three of them using more than 50 percent of their wages to buy anthracite. From 1819 to 1824 one of them ran up a coal bill of $74.59 for 68 tons, which he worked off in the ark yard. Accounts of the other carpenters indicate a similar pattern.[19]

John Smith needed shoes and leather gloves for his miners as well as leather halter straps for the horses. Between 1816 and 1822 he employed three shoemakers to make, mend, and sole shoes. One of these shoemakers took 62 bushels of coal for his work while another received his winter's fuel supply of 3.75 tons annually for three years in exchange for part of his bill.

Other skilled craftsmen worked for Smith and took their pay in coal. A millwright built Smith a dam for his gristmill and made the millstones. The total bill came to $220.98, and Smith paid one-third of the bill with 31 tons of anthracite delivered during three winters.[20] In similar fashion, from 1820 to 1824 Kingston's leading cabinetmaker secured a total of 22.5 tons of coal by making desks, bookcases, chairs, and a bedstead for John Smith. Even John Smith's son received some education in exchange for coal. During the winter of 1822–1823 Francis Smith obtained nine weeks of schooling for six tons of coal.

John Smith traded coal with local farmers for part of his food supply. In 1822 one farmer gave Smith 90 pounds of beef, the equivalent of a quarter steer, for 90 bushels of coal. This farmer purchased 36.75 tons of anthracite from 1819 to 1824 in exchange for agricultural produce. Another farmer sold Smith potatoes and other produce to pay for his coal during three years, and still another bartered corn for anthracite. Cider and whiskey were apparently the chief beverages of the Smith family and the miners. One Plymouth township resident earned his annual fuel supply during four years by making barrels of cider. The miners, of course, preferred whiskey to cider. From 1819 to 1824, one still near the Smith mine became the biggest local consumer of anthracite. John Smith sent two and three loads of coal a month to keep its fire burning. In July 1822 the still owner traded Smith 243 gallons of whiskey for 81 tons of anthracite, or three gallons of whiskey for one ton of coal. During this six-year period he received a total of 186 tons of hard coal.[21]

John Smith raised part of his food supply on a farm adjacent to his mine. Three of his regular employees spent some time helping to farm in addition to their work in the coal business. One of the men, who spent most of his time during the winter hauling coal, took 14 days in May 1822 to plow Smith's fields and garden. During the winter of

1816, John Smith had one of his employees help with the butchering. Each of these men received coal as a partial payment for his work.

John Smith's Plymouth store adjacent to the coal mine became the center of his operation. He kept the books at this store and made arrangements to trade his coal for other goods and services. The available account books, which deal only with the years 1815 to 1824, show the nature and scope of his business. During this decade he had a total of 85 different coal customers in Luzerne County. Several of these customers continuously purchased anthracite from Smith over four or five years and kept up their patronage after 1824.[22] Of course, the availability of anthracite at other mines reduced the average number of buyers each year.

Annual sales show the growth of Smith's business and indicate two different levels of consumption with the year 1820 being a turning point. For the first five years (1815–1819), Smith had an average of 15 customers annually who bought an average of 2.7 tons per customer. The value of his coal sales amounted to an average of $75 a year. During the next five years (1820–1824), the average number of annual customers jumped to 25, the total tons per year to 140, or 5.5 tons per customer, and sales to $170 per year.[23]

As discussed previously, Smith bartered almost all of this anthracite for goods and services. In this way he paid for most of his operating expenses. The local trade, which involved about 10 percent of his total coal production, took care of his labor bill, secured the services of blacksmiths, carpenters, leather workers, and shoemakers, and bought his food and drink. It sustained most of his business and allowed Smith to specialize in coal production. This same system kept the other Wyoming Valley merchants such as Jacob Cist in business and helped them grow even though their coal business was not as extensive.

Most of Smith's coal going down the Susquehanna River was used to finance the business and expand the trade. Evidence indicated a direct correlation between local consumption and the amount sent down river. From 1815 to 1819 John Smith shipped an average of six arks carrying about 360 tons annually to market. In 1820 he increased the number of arks to 10, and during the next four years an average of 20 of Smith's arks a year descended the Susquehanna, carrying about 1,200 tons to market each year.[24] Smith sold most of this anthracite on 90 days' credit. At $7 per ton Smith carried roughly $8,400 on his books for at least three months a year. Without the local coal trade to pay current bills, or other means of financial support, Smith could not have afforded to extend such a large sum on credit.

Smith's business must have paid off and permitted even further expansion. In 1828 a Plymouth merchant and coal dealer said that during the previous year and with a limited amount of capital his firm and that of a Kingston merchant had shipped a total of 10,000 tons of anthracite to market. He implied that several other Wyoming Valley concerns sent similar amounts to market.[25] Three other merchants in the valley, including George M. Hollenback and John Smith, had, by this date, sufficient capital and showed signs of each sending at least 5,000 tons to market annually.

If 25,000–30,000 tons of anthracite descended the Susquehanna River in 1827, Wyoming Valley coal production had kept pace with the amount going down the Lehigh and Schuylkill rivers. In 1827 the Lehigh Coal and Navigation Company shipped 32,000 tons to market, and the Schuylkill Navigation Company handled 31,000 tons.[26] While the Lehigh Coal and Navigation Company was the largest coal producer, two Schuylkill County companies marketed between 5,000 and 10,000 tons. Like most Wyoming firms, one-fifth of Schuylkill County's 70 miners produced 300 tons or less in 1827. Although small-scale enterprise dominated both the Schuylkill and Wyoming regions, only the bigger firms with capital to invest in the coal trade earned substantial profits. Merchants were very much aware of this fact. For merchants such as Jacob Cist, George M. Hollenback, and John Smith, selling or bartering anthracite at their stores had slowly developed into a good business.

As far as these merchants were concerned, the construction and operation of various types of mills followed storekeeping in importance. Even in this area of activity, there was a strong relationship between the growth of the anthracite trade and the operation of many mills in the Wyoming Valley. From 1815 to 1825 Jacob Cist's analysis of anthracite's impact on industrial growth started coming true. The use of anthracite in blacksmith shops and for making nails in Wilkes-Barre antedated Cist's arrival in the valley, and its use to make whiskey has already been noted. In June 1824 George M. Hollenback received word that the owner of a lime kiln was willing to exchange his product for anthracite at the rate of $1 per bushel of lime for $.50 per bushel of anthracite delivered.[27] In the 1830s Hollenback shipped several arks of hard coal annually to lime kilns in Columbia and Snyder counties.

Anthracite may have been used as early as 1818 to make bricks in Wilkes-Barre. In April 1818 George M. Hollenback hired three men to produce bricks at his brickyard. By the end of May the kiln began tempering clay, and two months later the construction of Hollenback's new brick store and house commenced. Some of the fuel was wood, but Hollenback most likely had the workmen try anthracite in the brick kiln. The success or failure of the experiment was not recorded, nor was the amount of fuel consumed. In 1820 George M. Hollenback moved into the new structure, which became his headquarters during subsequent decades.

By cutting boards for the coal arks, sawmills aided the growth of the industry. According to John Smith's accounts, one of the big 90-foot arks of 60-ton capacity required 2,000 feet of lumber. Profits could be made supplying the frames and boards for these arks. John Smith paid $20 per frame for his arks and $12 per 1,000 board feet of planking. Being the largest coal shipper in the valley, the Smith brothers bought a considerable amount of lumber each year. In May 1816 Abijah Smith placed an advertisement in the Wilkes-Barre newspaper for 20 ark frames and 30,000 board feet of two-inch white pine plank. He promised to pay cash on delivery.[28] The number of arks or ark frames they purchased each year was not recorded by John Smith. Surely this number corresponded with the number of arks of anthracite sent down the Susquehanna River annually.

In supplying the coal merchants with ark frames and planks, sawmill owners found a profitable enterprise, and they even engaged in constructing entire coal arks. In April 1817 Kingston's leading merchant and sawmill owner advertised two coal arks for sale. Hollenback and Cist entered into the business of supplying coal arks within a year after they built a sawmill on Mill Creek in 1820.[29] On 11 November 1821 Nanticoke's ironmaster and coal dealer sent a note to Hollenback's storekeeper stating that he could not get his coal ready for shipment in time to take advantage of the high water and therefore did not need any arks that fall. If anyone else wanted the arks, he continued, Hollenback should sell them. During the winter of 1821–1822, Cist and Hollenback planned to construct a number of coal arks, and in December they ordered 50 large ark ropes to be delivered by the following February. In 1822 Abijah Smith purchased coal arks from the Hollenback firm. A short time later he bought three arks, oars, and lining boards and went to Mill Creek to get them.[30] Hollenback's arks also took his own coal to market.

The number of arks descending the Susquehanna is almost impossible to determine. Three newspaper articles provide clues to the volume. In April 1804 a Kingston resident counted more than 1,500 arks, boats, and rafts which had gone downstream during the spring rafting season; most of them carrying agricultural products. No mention was made of coal.[31] In 1818 a Harrisburg resident made a similar count, but only for one day of the busy rafting season. If his calculations and prediction proved accurate, more than 2,000 river crafts went down the Susquehanna in 1818. Although most of these crafts still carried agricultural produce to market, anthracite had become an important item in the trade.[32] Most of these river crafts were broken apart and sold for wood at Port Deposit. The sawmills in the upper Susquehanna Valley operated constantly, cutting wood for new arks and boats to take anthracite and other products to market.

Compared with the gristmill, sawmill, and a large paper mill at Kingston which used anthracite, Hollenback's team and wagon cost very little. Yet this team and wagon were employed continuously, carrying merchandise for Hollenback and Cist. Most of the coal hauled by their wagon went from the mine to the riverbank. With the reduced price of coal after the war, Cist and Hollenback no longer found it profitable to transport anthracite overland from Wilkes-Barre to Philadelphia. On 8 March 1815 one of their men took one ton in a wagon to Germantown but sold only 21 bushels at $.38 per bushel. The $7.98 he received fell far short of the $20 for expenses, plus horse feed, incurred in the trip. The only time it paid to send anthracite from Wilkes-Barre to the Quaker City was when the wagon went to the city empty in order to pick up merchandise. Jacob Cist recorded only one such trip. In October 1818 his wagon master took one ton of hard coal to Philadelphia. The rest of the time his wagon carried flour and other agricultural products to the Quaker port.

As far as most Wyoming Valley merchants were concerned, developing new coal mines came at the bottom of their list of business activities. Only John Smith specialized in coal production. Keeping the account books and overseeing the business preoccupied Jacob Cist's time. These chores, however, failed to challenge his imagination. Scientific inquiry and plans to develop the anthracite industry offered an outlet for his energy. Cist recorded the nature and extent of coal mining in the northern field as he investigated the anthracite formation.

Between 1813 and 1816 Jacob Cist sketched the six best mines in the northern field. His picture of the Smith coal mine appeared in the *American Journal of Science* in 1822 (see figure 10). According to Cist, Smith had an 18-foot-thick vein of anthracite which dipped to the south at an angle of 20 degrees. Having a solid roof, this mine was worked in the pillar and breast fashion by leaving massive pillars about eight feet thick. By 1822 John Smith had driven six good sized slopes into the seam and had begun using black powder to loosen coal.[33] Four years earlier Smith had secured the services of a Connecticut stone-quarry blaster to teach him the skills of the trade and initiate the use of blasting powder. In October 1819 Dupont gunpowder was advertised in the Wilkes-Barre newspaper. It could be ordered directly from a Philadelphia merchant or through one of the Wyoming Valley tradesmen. Cist said that gunpowder was occasionally used in all the mines but was employed most extensively by Smith.[34] Blasting powder helped Smith produce the greatest amount of anthracite in the Wyoming Valley.

10. Smith's Coal Mine Drawn by Jacob Cist (*The American Journal of Science,* IV, 1822)

Jacob Cist said that almost all of the numerous mines in the Wyoming Valley were open quarries. In removing the superincumbent strata, some miners blasted coal off with gunpowder. But the most commonly accepted technique was to drill two-inch holes in a straight line from 12 to 24 inches apart. Into each of the holes miners dropped two long semicircular wedges. Long tapering wedges were driven between the semicircular wedges in an alternating pattern until a large mass of coal came loose. Miners then used sledges to break the large pieces into convenient sizes for handling.[35] In this manner anthracite was taken from the face of the open mines.

Matthias Hollenback used drills and wedges to mine his coal. His great mine was ideally suited for this technique. Cist said that Hollenback had an 18-foot seam which dipped at 6 degrees to the west and 12–15 degrees to the north. The strata remained exposed for a distance of 500 feet. The roof was 10–20 feet thick and fairly solid.[36]

Apparently, the roof of Hollenback's mine did not prove as solid as Cist believed. On 25 February 1823 a miner and his son were digging coal for Matthias Hollenback when the mine caved in and buried the boy alive. After a half hour of intensive labor, the father freed his son but not in time to save his life. The boy was so severely injured that he died the same evening, having remained senseless and speechless the entire time.[37] This was the first known death from a mining accident in the anthracite region.

Above this mine on Mill Creek, Hollenback owned another coal seam which Jacob Cist figured was 10 feet thick. Farther up Mill Creek was still another good mine with a seam 25 feet thick which dipped at an angle of only 12–15 degrees.[38] This slight angle would permit excellent mining.

Veins of coal six feet thick or more, which could be worked by standing fairly level, were considered the best potential mines. Except for the open quarry at Summit Hill, the northern field with thick veins and slight angles of inclination promised better mining conditions than the southern and two middle fields with their deep valleys and steeply pitched veins. Cist was aware of this fact.

The development of the Wyoming Valley coal trade evolved much more slowly than the Lehigh coal trade. The heavily capitalized Lehigh Coal and Navigation Company built facilities to rapidly increase its output while merchants in the Wyoming Valley used the methods proven effective in handling agricultural products to enlarge their coal trade. These policies helped establish a market for Pennsylvania anthracite in a region short of money.

6 Scientific Inquiry and the Coal Land Business

> His researches into the geological structure and formation of our portion of country, and particularly into the anthracite coal region of Pennsylvania, have been extensive and indefatigable; and while they have contributed to enrich the cabinets of many scientific men, both in this country and in Europe, with mineralogical specimens, they have also been a means of calling the attention of our citizens to those vast mines of combustible treasures with which our mountains abound, and which we trust will be the means, under Providence, of giving employment to thousands of industrious men, and prosperity and wealth to our country.
>
> Jacob Cist's obituary
> *Wyoming Herald,* Wilkes-Barre, 6 January 1826

Thus Jacob Cist's obituary paid tribute to one of the dominating themes of his busy life. Even before moving into the Wyoming Valley in 1808, he arrived at the conclusion that anthracite offered the greatest opportunity for exploitation and would become the leading sector in the area's developing economy.

Philadelphia's fuel crisis in 1813 led him in still another direction in establishing a market for Pennsylvania anthracite. Between the months when the fuel crisis first stirred public attention and 1816, Jacob Cist and Judge John B. Gibson, a graduate of Dickinson College who shared Cist's interest in geology, explored the northern anthracite formation.[1] In addition to sketches of Wyoming Valley mines, Cist made three detailed drawings of mines in the Lackawanna Valley. All of these drawings indicated the size of the coal seams and provided other information necessary to attract investments into the region to help develop the anthracite trade.

The mine he drew at Pittston had a six-foot seam with a long face of coal exposed to view and covered three tracts of land. Like others in the Wyoming Valley, it could be easily worked by taking off the top cover of slate and earth. The mine lay 100 feet above the river, and the terrain permitted easy shipment to the riverbank. Another mine drawn by Cist was four miles up the Lackawanna Valley on Spring Brook near present-day Scranton. It had an 18-foot face and was situated so that coal could be delivered to boats in the river at little expense.[2]

A Lackawanna Valley mine sketched by Cist became one of the most famous in the northern field. It was located at present-day Carbondale and belonged to two Philadelphia merchants, Maurice and William Wurts. Cist noted that a 30-foot seam of high-quality anthracite was visible for a distance of 300 feet at the Lackawanna Creek. The stream flowed across the bed. The Wurts brothers had erected a crude pump to keep the mine dry and had taken out a few hundred tons of anthracite. Cist noted a coal pile amounting to about 300 tons alongside the river.[3]

Cist's drawing of the famous Lehigh coal mine at Summit Hill also falls under the category of scientific investigation in order to attract capital investments. He indicated accurately that the mine was the outcrop of a mammoth vein which folded as it followed the contour of the ridge, permitting a large quantity of anthracite to be quarried quite easily. Cist said that the mine contained two seams, one 50 feet thick and the other 9 feet thick.[4]

Note also that Cist drew a 10- to 15-foot thick vein of red ash on the diagram. He ranks among the first men to call attention to the distinction between red ash and white ash anthracite, one of the prominent features characterizing the two basic types of Pennsylvania's hard coal. After anthracite burned, the remaining ashes appeared either red, indicating the presence of iron pyrites and sulfur, or white, signifying pure coal of the highest carbon content. In his scientific notes Cist said that the lowest veins, which formed the floor of the coal formation, contained pyrites, and the upper veins were more nearly pure carbon or white ash. These facts, which he observed both at the Lehigh mine and in the Wyoming Valley, were true for the entire formation.[5] By carefully studying the composition of different coal specimens, noting the relationship of different types of veins, and making simple logical deductions in light of information known about the British coal formation, Cist concluded that the anthracite formation was composed of these two types of anthracite. His conclusion proved to be accurate.

Jacob Cist and John B. Gibson paid the greatest attention to the coal measures in the Wyoming Valley. From their Wilkes-Barre homes in the center of the valley they could easily travel to nearby mines and outcrops in order to make sketches and collect specimens. Over two-thirds of Cist's drawings focus on the Wyoming Valley coal formation. His conclusions about this formation were tested as he traveled to more remote regions. Cist's careful observation of natural phenomena enabled him to arrive at accurate conclusions decades before others interested in anthracite noted the same facts.

Cist observed that the entire anthracite formation of Pennsylvania rested on one of red sandstone which became known as Mauch Chunk red shale.[6] As they moved along dirt roads and footpaths, travelers easily spotted this red shale. It served as a distinctive feature of the region, separating the coal measures from the brown sandstone and clay where farmers cultivated their fields. To the trained eye, this red shale was a sign that a person was either approaching or leaving the anthracite formation. A band of red shale completely encircled all four fields. Other anthracite experts drew the same conclusion as Cist. In 1836 a Pottsville mining expert made the first map of the southern and two middle fields which showed individual tracts and their location to prominent geographical features. Like subsequent map makers, he carefully marked off the anthracite formation by showing the position of red shale.[7]

To determine the size and shape of the northern coal field, Cist traced out the strata which formed its outer circumference. He located three veins which outlined the western end of the formation. The outer vein, averaging about 24 feet in thickness, was the great red ash vein. A second vein was about 3 feet thick, and a third varied from 6 to 8 feet thick.[8] Following these veins from Shickshinny eastward to the mouth of the Lackawanna River, Cist accurately mapped the shape of the formation in his district (see figure 11).

From his knowledge of the Lackawanna Valley, Cist figured that these seams continued for another 26 miles, terminating in Susquehanna County near present-day Forest City. As figure 12 indicates, Cist gave fairly accurate measurements for the length of the three districts in his area: 26 miles for the Lackawanna, 18 miles for the Wyoming Valley, and 5 miles for Newport. He made three logical divisions for the northern field, each of which became a major mining district in the nineteenth century. Cist said the formation was three to four miles wide.[9]

Cist's sketch of the Luzerne formation showed nine coal seams. He determined the number of veins by carefully studying a cross section of the Wyoming Valley one mile east of Nanticoke Gap. At this point barren cliffs permitted examination of the rock formation. He calculated these dimensions of the nine coal veins: the 24-foot red ash vein

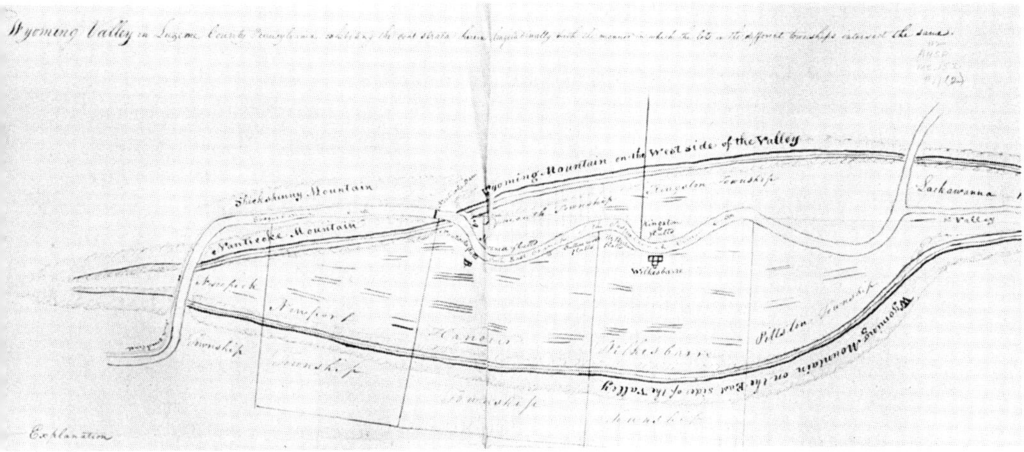

11. The Wyoming Valley in Luzerne County, Pennsylvania, Exhibiting the Coal Strata, Drawn by Jacob Cist (Courtesy Academy of Natural Sciences of Philadelphia)

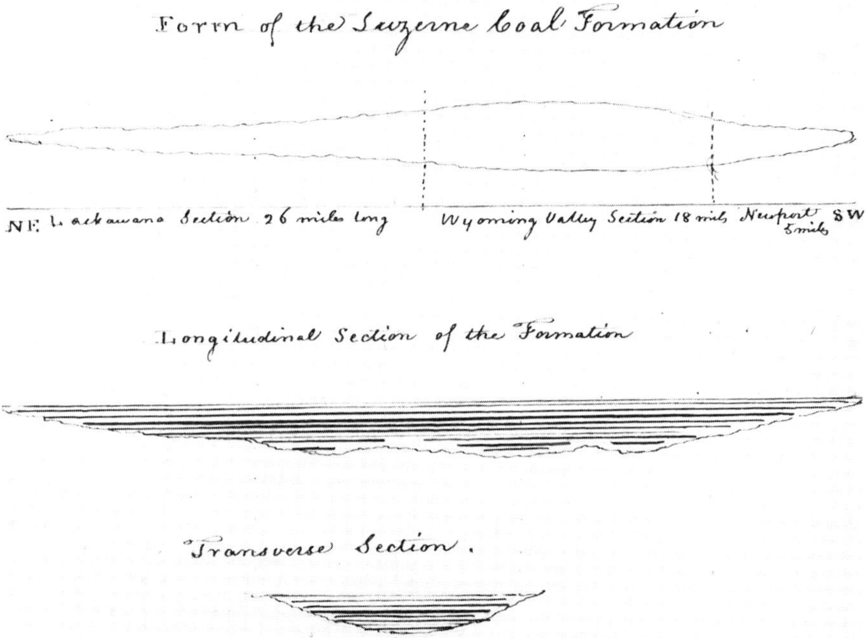

12. Form of the Luzerne Coal Formation, Drawn by Jacob Cist (Wyoming Historical and Geological Society)

and the 3- and 6-foot veins which transversed the entire valley, as well as six veins of 4, 6, 8, 3, 9, and 3 feet.[10] According to the most accurate examination of this area made at the end of the nineteenth century by the second geological survey, Cist's calculations were exceptionally good.[11]

Jacob Cist's collection of notes and drawings on the anthracite formation is the earliest and most accurate known to exist. He covered every major point which would later be considered important by mining engineers: the size and shape of the formation, the number and thickness of strata, their angle of inclination, other minerals intermixed with the coal, the relationship of particular mines to other geographical features, and the number of tons in a tract as well as the entire formation. This information was important to investors. Between 1813 and 1816 Cist spent many hours making drawings of the formation and taking notes on different geological strata. The time and energy he spent taking notes on the geological formation, locating coal seams, and sketching pictures of potential mines went beyond Cist's love for drawing and his curiosity concerning natural science even though his ability to draw and his interest in science gave him personal satisfaction. In his correspondence with close friends and business associates, Cist candidly admitted his desire to make money by developing the anthracite trade.

One of the chief ways for Cist to make money was by selling land which Matthias Hollenback and others had acquired prior to 1800. If Cist could prove that his county contained a valuable national resource, real estate in his region would sell above the national average. His task would not be easy because at this time cheap fertile land existed west of the Appalachian Mountains, and businessmen in all parts of the country were competing for the scarce capital owned by merchants and financiers in the Atlantic coastal ports in order to develop their respective regions.[12] By mapping the coal formation of northeastern Pennsylvania, Cist played a vital role in convincing businessmen about the value and abundance of anthracite and consequently attracted investment into a region short of capital.

Cist's work was part of a general movement going on throughout the United States. In 1818 Benjamin Silliman, professor of Chemistry and Mineralogy at Yale College, established the *American Journal of Science*. He intended the journal to include all the physical sciences as well as their application to the arts and every useful purpose. He wanted original contributions concerning every branch of science. He singled out the need for more knowledge about American mineralogy and geology, including descriptions and maps of particular states and districts.[13] Silliman achieved his objective. Starting with the first issue and continuing thereafter, the journal contained articles describing various minerals, their practical application, and the geographical areas from which they came. Within four years the *American Journal of Science* became a successful commercial venture and a respected scholarly publication.

Jacob Cist read the journal and responded to one of its inquiries. In the first volume Alexandre Brongniart, a member of the Royal Academy of Sciences and Engineer Mines in France, who was well known in scientific circles for his work in determining the age of geological strata by using fossils, requested help in securing specimens. He was working on a large-scale synthesis of rock stratification, comparing Europe with the United States, and he wanted to collect specimens of fossils and the accompanying rocks. In particular, he was looking for distinguishable impressions and remains from coal countries.[14] Silliman volunteered to handle the transfer of American specimens to the French scientist.

On 24 July 1821 Cist sent a collection of Wyoming Valley rocks, coal specimens, and fossils to Brongniart via the Yale professor. Accompanying this collection was an account of the anthracite mines in the region around Wilkes-Barre and Cist's pamphlet advertising the advantages of hard coal. Most of the information which Cist had collected in his notes appeared in the account. As a result, Silliman honored Cist by publishing his account and pamphlet as the lead article in the next volume of the journal.[15] Silliman also sent the information to Brongniart, who acknowledged the valuable aid rendered by Cist. As a matter of fact, Brongniart affixed the scientific name "cistii" to a number of anthracite specimens in honor of the Wilkes-Barre merchant.[16]

Cist contributed two other articles to the *American Journal of Science*. On 10 May 1821 he sent Silliman samples of manganese found along several creeks in the Wyoming Valley and on Broad Mountain. Noting that manganese had brought high prices during the War of 1812, Cist felt that its location and qualities should be brought to public attention. He said this mineral was essential to the chemist, bleacher, and potter. Three years later Cist gave Silliman several specimens of vegetable remains found in the anthracite deposits. He felt that these samples showed that hard coal was formed from vegetable matter. Silliman noted that the specimens were uncommonly fine and praised the Wilkes-Barre merchant for helping enrich his scientific cabinet. He hoped that others, living near coal mines, would make the effort to collect similar specimens and donate them to his collection.[17]

Such specimens were important to students of mineralogy. Although it was impossible for them to visit every area containing valuable minerals, a wide variety of minerals could be gathered inexpensively in scientific cabinets and compared with similar specimens from other areas of the world. By subjecting small samples to heat and acid tests, students could learn the qualities of various minerals. As early as 1805 James Woodhouse, a professor of chemistry at the University of Pennsylvania, tested Lehigh coal. His results became fairly well known to other American scientists. Similar tests were made by other chemists throughout the nineteenth century. Cist included this information in his general accounts where it could be compared with results of students and scientists doing original research.

Besides Benjamin Silliman and Alexandre Brongniart, Pennsylvania anthracite was attracting the attention of other scientists. Thomas Cooper introduced the subject of anthracite into his lectures on mineralogy while he taught at Dickinson College, the University of Pennsylvania, and South Carolina College (now the University of South Carolina). In January 1819 Cooper offered a course in Philadelphia on mineralogy, which he advertised in area newspapers. Again Pennsylvania anthracite came under consideration in his lecture.[18]

Several other American men of science turned to Jacob Cist to enrich their information about anthracite. In April 1821 Cist's friend John B. Gibson, who had become an associate justice of the Pennsylvania Supreme Court and a director of the American Philosophical Society, wrote to the Wilkes-Barre merchant about visiting him in Philadelphia. He wanted to see more of Cist's drawings of the fossil relics found in the hard coal formation. Gibson told Cist that he would receive full credit for all the information he contributed. Gibson noted that the subject was a new one and by "no means devoid of interest. Everything is so which respects the geology of this continent." This jurist, who made reports about geology to the American Philosophical Society, had become interested in anthracite while living in Luzerne County.[19]

In August 1824 James Renwick, professor of natural philosophy and experimental chemistry at Columbia College (now Columbia University), received a collection of anthracite specimens from Cist. Renwick was so pleased with the gift that he sent the Wilkes-Barre merchant a box of minerals. Among the items it contained were coal metals from the great Scottish coal field. The Columbia professor felt that Cist would find considerable satisfaction in comparing these specimens with similar stones found in Luzerne County.[20]

Another of Cist's correspondents was Thomas Say, curator of the American Philosophical Society and professor of natural history at the University of Pennsylvania. On 11 December 1824 Say wrote to Cist about one of his friends who was interested in introducing Susquehanna coal into general use. Say's friend had called at the American Philosophical Society for the purpose of perusing a letter in which Cist had made some very interesting calculations concerning the probable quantity of anthracite in Luzerne County above water level. Doubts existed in some minds as to the accuracy of the figures. If it was not too much trouble, Say asked Cist to send him the figures for Luzerne County or inform him where they could be found.

Cist sent Thomas Say the requested information, stating that the Wyoming Valley contained nearly three billion tons of anthracite. He repeated other points mentioned previously in this chapter. The following year Cist sent the London Geological Society a similar statement, including a map of the entire hard coal formation and a table showing the nine strata (see figure 13). Cist estimated that the Pennsylvania formation was 120 miles long and contained eight to ten billion tons of coal. A more detailed study of the map (figure 14) revealed that it was accurate in many ways, and resembled others of the hard coal formation (see figure 15).

From 1815 to 1825 Jacob Cist became the leading expert on the Pennsylvania anthracite formation. As interest in American geology increased, so too did the demand for Cist's knowledge. His information reached scientific circles throughout the United States, England, and France. Four leading scientists, Brongniart, Renwick, Say, and Silliman, recognized his contribution as being significant.

Cist's reputation as an individual who understood minerals and their practical value extended beyond the scientific community. As noted in Thomas Say's correspondence, businessmen were aware of Cist's work. Nicholas Biddle wrote to Cist for information about minerals located on one of his properties in the upper Susquehanna Valley. He told Cist that his reputation as a gentleman of science and character had induced the Philadelphia banker to take the liberty of initiating the inquiry. Biddle wanted to know about the contents of salt in a well near Montrose and asked Cist to look into the matter.[21] Biddle did not mention Wyoming Valley coal, perhaps because he had already turned his attention toward developing the Schuylkill coal trade.

Jacob Cist served as one of the principal agents in Luzerne County, helping financiers determine the value of their land. His scientific knowledge of the anthracite formation came in handy in making appraisals. In December 1819 a Northumberland, Pennsylvania, land agent called upon Cist to investigate a piece of property lying in the upper end of the Lackawanna Valley at present-day Carbondale. The land belonged to Lord William Russell, a Birmingham, England, merchant.[22] As Luzerne County tax collector from 1816 to 1818, Cist no doubt had been in contact with the Northumberland County land agent, who had learned to respect his ability and judgment.

Cist explored the land and made a map of Russell's Lackawanna Valley estate showing 22 tracts of nearly 10,000 acres.[23] On the map

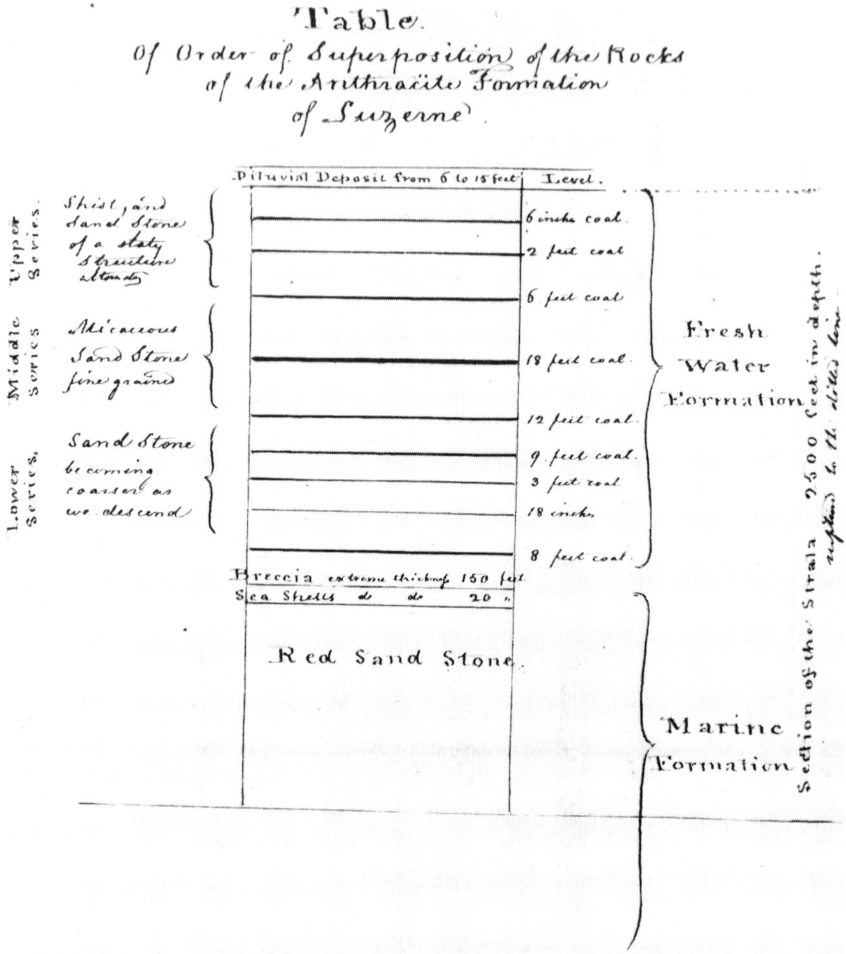

13. Order of Superposition of the Rocks of the Anthracite Formation of Luzerne, Drawn by Jacob Cist (Wyoming Historical and Geological Society)

appeared the name of each tract, the number of acres it contained, and its location in relation to other tracts, the Lackawanna River, and county line. A description of the land noted the distances to the Delaware and Susquehanna rivers, the number of squatters, and the existence of several veins of stone coal along the banks of the Lackawanna River.[24] This estate became the property of the Delaware and Hudson Canal Company.

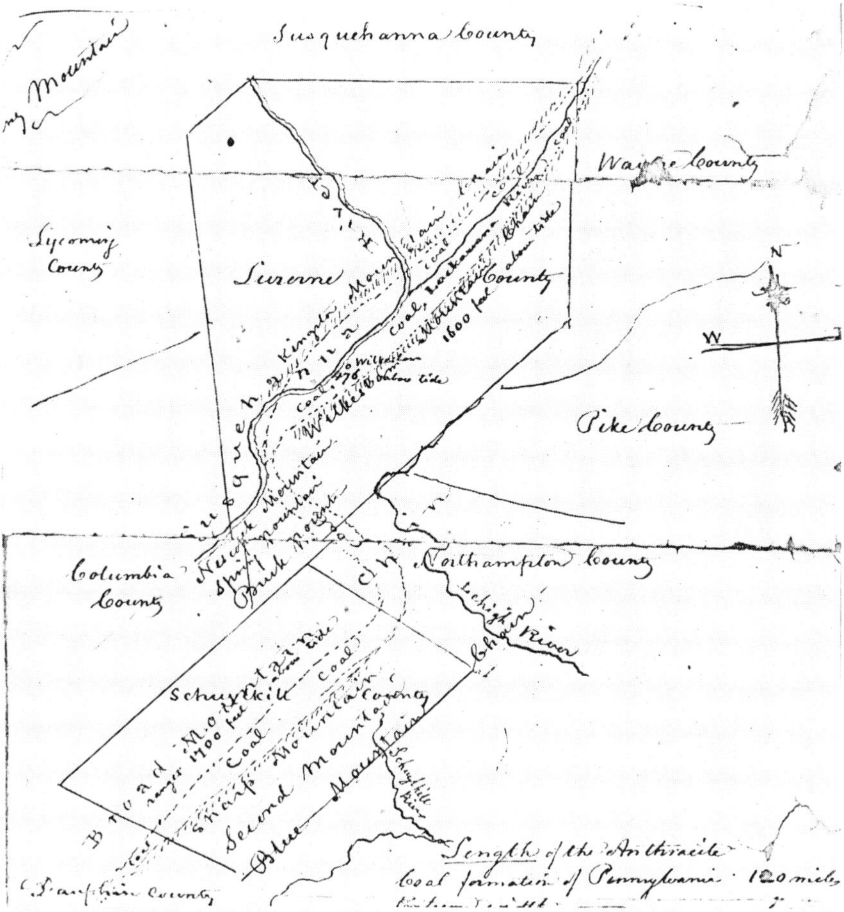

14. Form of the Anthracite Formation of Pennsylvania, Drawn by Jacob Cist (Academy of Natural Sciences of Philadelphia)

Making such surveys permitted Cist and others to gain firsthand knowledge of the land in their immediate vicinity. Surveyors took note of the types of soil and minerals appearing on the surface as they ran lines and staked out tracts. It was as important to learn what natural resources the land contained as the exact dimensions and location of the property. The surveyors of northeastern Pennsylvania hunted for veins of anthracite coal, anticipating a rise in land values if they could spot good coal seams.

From the Luzerne County surveyors and his own fieldwork, Cist

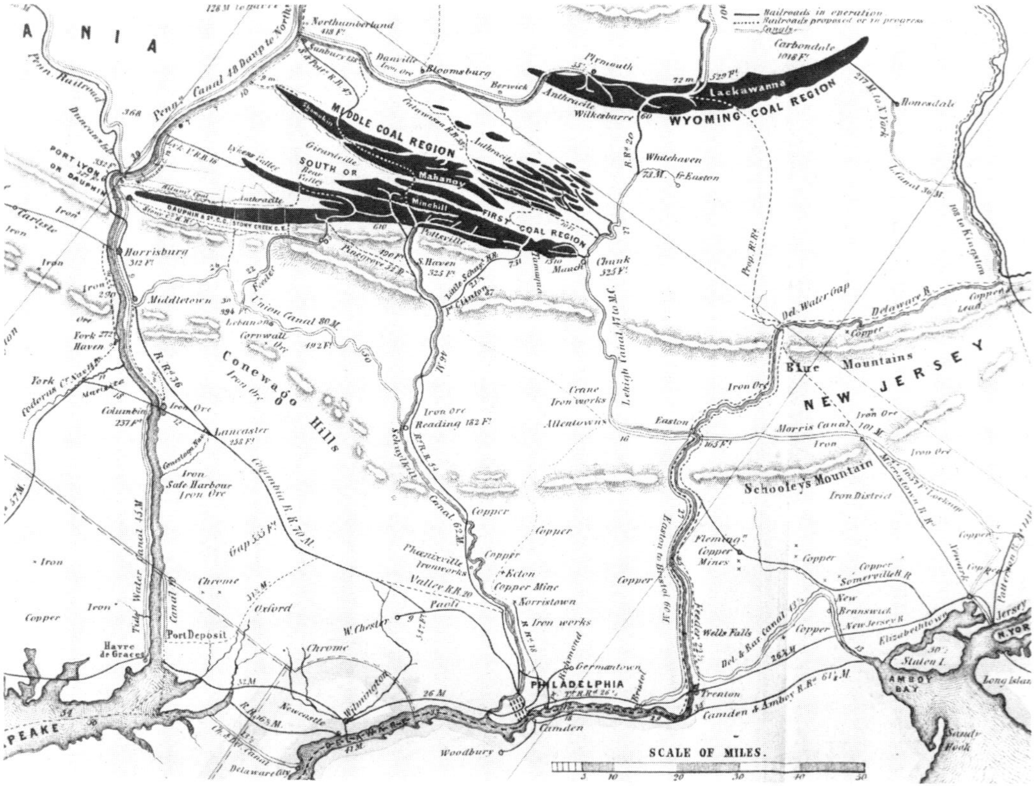

15. "Map Illustrative of the Statistics of the Coal Trade of Pennsylvania," in Richard Cowling Taylor's *Statistics of Coal* (Philadelphia: J. W. Moore, 1848)

kept tabs on the location and dimensions of the northern anthracite formation. He had access to the information concerning Matthias Hollenback's land which stretched from the southern part of the county north to the New York State line.[25] Cist used this information in determining the value of land in the northern coal field.

Philadelphia businessmen were interested in promoting the Susquehanna anthracite trade and drew upon Cist's notes to further their cause. In September 1825 a group of Philadelphians visited Luzerne County and talked with Jacob Cist. They had been impressed with what Cist had said at the Harrisburg canal convention the previous month and wanted to see the geological formation firsthand. During their visit Cist showed them the coal seams and geographical terrain. In examining the formation, they had the chance to see the various features of the area which Cist had elucidated in his notes. They believed Cist's statements were accurate and became engrossed in his dream to develop the anthracite trade.

Before supporting transportation projects and purchasing coal land, investors turned to Jacob Cist and others who possessed some scientific knowledge of the geological formation and knew the local terrain. Businessmen wanted to know whether the land contained thick seams of anthracite for long-term mining, good outcrops for quick and easy exploitation, and other geographical advantages for getting coal to market. Only men with the type of scientific knowledge which Cist had accumulated could provide sound advice about a potentially profitable project. Like Nicholas Biddle, these investors depended upon individuals whose reputations as gentlemen of science and character were well known.

Jacob Cist was not the only person in northeastern Pennsylvania to earn such a reputation. His extended family as well as his close friend Isaac Chapman were looked to with similar respect. In July 1816 Chapman announced the opening of a land office in Wilkes-Barre and set forth the terms under which he was willing to do business. He claimed to have extensive knowledge of land in Luzerne, Bradford, and Susquehanna counties.[26] Chapman became one of the most successful land agents in the area during the next decade, and his clients numbered among the largest landholders as well as the Commonwealth of Pennsylvania. To aid businessmen, in 1816 the Pennsylvania legislature passed a bill for surveying every county to determine township lines, geographical features, and mineral deposits. Isaac A. Chapman was commissioned to make the maps for Luzerne and Northampton counties. His area included all of the northern field, middle eastern field, and eastern end of the southern field. His wartime experience in the Lehigh region helped him recognize geographical features essential for developing the anthracite trade, particularly the sites for potentially good coal mines. Along with Cist and Matthias Hollenback, Chapman possessed the greatest overall knowledge of geography and the coal formation in northeastern Pennsylvania.

Chapman marked major outcrops in the northern field on his Luzerne County map. He located four outcrops along the road from Nescopeck to Wilkes-Barre near the present-day towns of Glen Lyon, Wanamie, Nanticoke, and Ashley. John Smith's mine at Plymouth received an *x* as did Hollenback's Mill Creek mine and those at the base of the mountain south of Wilkes-Barre. Chapman noted still other coal deposits just east of Pittston and near present-day Scranton.[27] These points included the drawings of coal mines made by Cist.

On the 1817 Northampton County map Chapman located stone coal at Beaver Meadows, Room Run near Nesquehoning, and Summit Hill.[28] His failure to draw attention to anthracite outcrops in the Hazleton region, Black Creek Valley, or upper end of the Lackawanna Valley may have rested on his lack of knowledge about coal in these areas. But more likely it was due to the few settlers in these places and the geographical obstacles in getting this coal to market, since Chapman clearly recognized signs indicating the existence of coal and believed in the potential use of this fuel. He may also have been biased toward his friends and associates who already had opened coal mines and felt that the deposits in these remote areas could not be brought into profitable production in the foreseeable future. Chapman's maps are the earliest ones to show the precise location of numerous outcrops in the northern and middle eastern fields. During the nineteenth century, major mining towns developed adjacent to these outcrops.

Chapman worked for many of Philadelphia's leading merchants, such as Henry Clymer and Samuel Mifflin, and became one of the agents for the North American Land Company. Altogether the amount of land under his care exceeded 130,000 acres. Of course, not all of this land was within the coal measures, nor did most of his work involve hunting for coal mines and determining the value of coal land. Chapman spent most of his time surveying land, paying taxes, and taking care of the land transfer to local settlers who were mainly farmers.

As an agent for the North American Land Company, searching for coal was one of Chapman's specific duties. In 1821 he began the task of finding coal on this company's land. The next year he explored Laurel Creek and Buck Run, tributaries of the Lehigh River, and marked off 12 tracts of land which a decade later became the estate of the Buck Mountain Coal Company, the second company to produce anthracite in the middle eastern field. In the 1830s five leading Philadelphia financiers including Thomas F. Wharton organized the Buck Mountain Coal Company.[29] The initial plans for such a company began in 1820 when a group of financiers employed an experienced miner who operated the Beaver Meadows mine to examine the property. He made a map of the terrain and marked a coal pit at the head of Buck Run which had produced some coal.[30] Chapman confirmed his findings.

A direct link existed between scientific inquiry and the establishment of a market for Pennsylvania anthracite. Before businessmen would pour huge sums into costly transportation projects, land, and mines, they wanted accurate geological and geographical information about the anthracite formation. Creating this body of knowledge was part of the work of Jacob Cist and others. With reputations as gentlemen of science and character, they were trusted to give sound advice. Major American merchants who invested in banking, commerce, manufacturing, and mining were far too diversified and too busy to know all the details about one enterprise. They relied on local experts to judge the practicality and profitability of specific projects. Without scientific knowledge about geology and geography the judgment of men such as Cist would have been faulty.

7 Anthracite and Pennsylvania's Internal Improvements

Jacob Cist recognized that the final task in establishing a market for Wyoming Valley anthracite concerned the construction of adequate transportation down the Susquehanna Valley so that a steady supply of hard coal from his region could reach Philadelphia and the other Atlantic coastal cities from early spring to late fall instead of during the short rafting season. However, the opportunity to carry out his plan did not arrive until 1825. To acquire a steady supply of anthracite, Philadelphia capitalists, during the fuel crisis of 1813–1814, had agreed to back construction of the Schuylkill navigation system. The Schuylkill Navigation Company, led by Cadwalader Evans, Jr., and backed by the financial resources of Stephen Girard, did not complete its project until 1825.[1] Meanwhile, New York State undertook construction of the Erie Canal to acquire the agricultural trade of the West. The instant success of the Erie Canal in 1825 threatened Philadelphia's western trade and forced Pennsylvanians to consider the necessity of constructing a transportation system from the Quaker port to Pittsburgh and Lake Erie.

During the first week of August 1825, 114 delegates from throughout the Commonwealth of Pennsylvania met in Harrisburg to discuss the need for better transportation facilities, and particularly for a canal or railroad from Philadelphia to Lake Erie via Pittsburgh. Almost every county sent at least one representative to voice the claims of his particular region. Jacob Cist and Nathan Beach, who owned coal mines at Beaver Meadows and Shickshinny, represented Luzerne County. At this meeting Cist put forth certain ideas which eventually led to compromise among various factions in the state and became Pennsylvania's transportation policy.

Wyoming Valley merchants had long recognized the need for improving transportation in their area. Shipping goods down the Susquehanna River on rafts during the high water season was far too risky and dangerous even for skilled boatmen, and the rafting season was too short a time period to suit their commercial needs. These merchants desired to improve river navigation or construct canals so that merchandise could more easily be sent north into New York State and south to Baltimore and Philadelphia. They were also interested in improving the Lehigh River as well as roads across the mountain leading to Philadelphia.[2]

In the Wyoming Valley serious discussion of navigation problems commenced in 1818, the year in which New York started work on the Erie Canal, Josiah White turned his attention from Schuylkill navigation to the Lehigh coal trade, and the Wyoming Valley coal trade began to develop more rapidly. On 9 January 1818, a Wilkes-Barre newspaper contained an article, most likely written by Jacob Cist, about transportation on the Susquehanna and Lehigh rivers. Basing his article on information received from Matthias Hollenback and three experienced rivermen, the author described the Susquehanna River. He listed the names of each rapid, the distance from one rapid to the next, the length of each rapid, and the cost of improving the channel at each rapid from Wilkes-Barre north for a distance of 103 miles to the New York State line. The channel could be greatly improved with $5,000. In addition to helping Wilkes-Barre merchants, this section would become part of a lake-to-sea route from Philadelphia up the Lehigh River to the Luzerne County seat and up the Susquehanna via the Finger Lakes to the Great Lakes. As the article commented on improving Lehigh River navigation from Stoddardsville, 11 miles south of Wilkes-Barre, to Easton, it claimed that "Luzerne County can furnish the citizens of Philadelphia with coal for fuel cheaper than it can be procured any other place."[3]

The next issue of the *Susquehanna Democrat* contained a report of commissioners who had examined the Susquehanna River to recommend major improvements. The commissioners called for spending over $20,000 for the entire river with $6,900 for the North Branch from Northumberland to Tioga Point. If the Pennsylvania legislature appropriated $5,000 annually for the next three years, the most dangerous rocks could be removed from the river bed.[4]

According to a Harrisburg newspaper, river traffic the following spring was heavy, justifying the cost of removing boulders. One day in May, 86 rafts, 16 arks, and 12 boats floated past the state capital. This volume of traffic would continue for more than 20 days, noted the river observer.[5]

Periodically, articles appeared in the local newspapers calling public attention to the need for better transportation facilities. Each year, when the state legislature met, the effort to acquire funds for river improvements was renewed. On 1 January 1819 the *Susquehanna Democrat* said that a considerable portion of Pennsylvania's population lived along the banks of the Susquehanna and its tributaries. The article noted rich farm lands, the growth of towns, and the inexhaustible mines of anthracite which had been discovered and opened. After citing four legislative acts for improving river navigation and the amount appropriated, the paper asked if this money had ever been spent.[6] The journalist desired more funds for internal improvements.

The initial success of Josiah White and the Schuylkill Navigation Company attracted attention in Wilkes-Barre. In February 1821 the *Susquehanna Democrat* reported that Schuylkill coal had been carried by sleigh 34 miles from the mines to Reading and was selling for $.18 per bushel instead of the usual $.45–.50 per bushel. The next year another news item noted the arrival of over 4,000 bushels of hard coal in Philadelphia by means of the Schuylkill navigation system, and more arks of Schuylkill anthracite were expected soon.[7] A December 1821 article in this Wilkes-Barre paper compared the price of Lehigh with Liverpool coal. It used material from an article by the editor of Philadelphia's *Democratic Press* who calculated that in New York City Liverpool coal cost $.43 per bushel and Lehigh coal $.30 per bushel, giving Lehigh anthracite an advantage of $.13 per bushel. Contending that 1.5 bushels of Lehigh coal equaled 2 bushels of Liverpool coal, the Philadelphia editor claimed that Lehigh anthracite had the overwhelming advantage as far as price and quality were concerned.[8]

The author of the article in the *Susquehanna Democrat* felt that the Philadelphia editor's calculations were correct and went on to say that the same quality of anthracite could be found near Wilkes-Barre. Like Cist, he maintained that this hard coal was at least 25 percent better than Liverpool coal. Furthermore, he stated, by means of the Susquehanna Valley, the Wyoming Valley could supply New York and other seaboard cities with coal as cheaply or cheaper than they could be supplied from any other place. He was not acquainted with the cost of water carriage from Havre de Grace to New York City but thought that it would not exceed $2 per ton. Since Wilkes-Barre coal could be delivered at Havre de Grace for about $6–7 per ton, he believed that anthracite from the Wyoming Valley could be delivered as cheaply in New York City as coal coming from the Lehigh Valley or Liverpool. Luzerne County coal producers could compete effectively in the eastern urban market.

This journalist proposed forming a company with a capital of $30,000–50,000 to handle the Wyoming Valley coal trade. The money would be employed mainly in digging coal and conveying it to Havre de Grace where a coal yard would be established for its reception. The company would own schooners carrying about 100–150 tons of anthracite constantly from Havre de Grace to New York, New London, Boston and elsewhere. In each city it would set up coal yards and keep them supplied. If the company kept a few properly constructed grates burning at each yard, Wyoming Valley anthracite would attract public attention and recommend itself as a valuable fuel. The author urged financiers to consider his plan, claiming it would greatly benefit the Wyoming Valley, all the counties of Pennsylvania and Maryland bordering the Susquehanna River and the Chesapeake Bay, and the cities on the Atlantic seaboard. He felt that the sooner his project reached fruition the better.[9] This plan resembles other proposals made by Jacob Cist, and the article may have been written by the Wilkes-Barre merchant.

In May 1822 the expected arrival of 200 tons of John Smith's coal in Philadelphia was noted by the *Susquehanna Democrat,* which hoped to see it advertised fairly soon in Quaker City newspapers. Three months later stone coal again came to Wilkes-Barre's attention when the local paper pointed out that the Susquehanna River served as the natural outlet for Wyoming Valley anthracite to reach market. Repeating a point that was previously made, the journalist said that he had long been of the opinion that New York, Boston, and other Atlantic coastal cities could be supplied with Wyoming Valley anthracite cheaper than if they received it from the Lehigh or Schuylkill valley. Luzerne County merchants, however, lacked capital to carry on the trade extensively. Looking to Baltimore financiers for aid in achieving his proposal, the journalist asked two Baltimore editors if they would be kind enough to find out the cost per ton of shipping coal from Havre de Grace to New York or Boston.[10]

In the same issue of 2 August 1822, the *Susquehanna Democrat* announced the removal of rocks from one particularly dangerous rapid where six arks had been wrecked during the previous boating season. This point had always been considered one of the worst, if not the very worst, in the descending navigation. Boatmen would be happy to learn that henceforth passing this rapid would be much safer and easier.

The Wilkes-Barre newspaper said that the river needed still further improvements and went on to show the value of commerce during the spring 1822 boating season. The Baltimore *American* had published a recent article reporting statistics from a gentleman at Port Deposit who kept a register of all produce descending the Susquehanna to that place. Between the opening of the river for navigation and 16 June 1822, he counted 487 arks carrying various goods valued at over $1 million. His figures did not include the lumber rafts which passed Port

Deposit. The Baltimore *American* felt that during 1822, between $1.5 and $2 million worth of goods would descend the Susquehanna. This much commerce merited additional expenditures to improve navigation.[11] Thus Luzerne County residents increased public pressure to improve Susquehanna River navigation by noting that the amounts spent on the Erie Canal and the Schuylkill navigation system, both near completion, were far greater than that spent on their river.[12] County merchants believed that the area they serviced held as much potential for good farming and mining as the Genesee or Schuylkill valley.

As the next state legislative session approached, Wilkes-Barre merchants called a public meeting to be held at the Luzerne County courthouse on 4 November. Matthias Hollenback chaired this meeting of leading county merchants and politicians. Noting that the population of the counties watered by the Susquehanna had grown from 278,000 in 1810 to 378,000 in 1820, they reviewed the natural resources which had attracted people to central Pennsylvania. The region, they believed, had the potential of sending $5 million to market annually if river improvements received adequate appropriations.[13]

The Luzerne County meeting created three committees to achieve its goal: a committee to draft petitions, a committee to secure signatures for the petitions, and a general committee of correspondence. Membership on these committees included one of Wilkes-Barre's leading lawyers and also the editor of the *Susquehanna Democrat*. The 18-man committee to acquire signatures had representatives from every township and counted among its members Nathan Beach, Isaac Chapman, and John Smith. The Plymouth coal merchant also served on the 10-man correspondence committee along with other coal merchants from Kingston, Nanticoke, Pittston, and Wilkes-Barre. Wilkes-Barre's most influential lawyer chaired the correspondence committee, and Jacob Cist did most of the work.

In addition to writing several articles for the Wilkes-Barre newspaper, on 5 December 1822, Cist wrote a letter to the Baltimore *American* which reflected the sentiments of his committee. After reviewing the nature of navigation of the Susquehanna and the goods sent to market, Cist said that the one product with the greatest potential for increasing trade was anthracite. Scarcely 5,000 tons of hard coal descended the Susquehanna annually because of poor navigation. However, if the river was improved for boats carrying 25–30 tons, the same size as those used on the Schuylkill River, Baltimore would receive enough anthracite to rival Philadelphia in the coal trade. Cist estimated that Great Britain consumed annually from 2 to 3 million tons and said that Luzerne County could supply a similar quantity to the port of Baltimore. His geological explorations had convinced him of this fact.

The amount of anthracite descending the Susquehanna River "alone [would be] sufficient to warrant the expense of rendering the river completely navigable."[14]

Luzerne County residents petitioned the Pennsylvania legislature during every session from 1822 to 1825 for river improvements. While many reasons were given to support the project, one point they always emphasized was the need for a canal-slack water navigation starting at the mouth of the Susquehanna and crossing Pennsylvania to the Great Lakes. Their proposed route followed the Susquehanna River north from Harrisburg up the North Branch, connecting with Seneca Lake and ultimately Lake Erie. In Wilkes-Barre this project became known as the National Route.[15]

A public meeting in Wilkes-Barre on 12 February 1825 listened to a discussion of the merits of six different proposed routes from the Atlantic to the Great Lakes. According to a statement prepared by Cist and Judge David Scott, the distance from New York City to Lake Erie via the Erie Canal was 520 miles with a height of 655 feet to overcome by locks. In comparison, the National Route was 511 miles and would require 1,593 feet of lockage. One route from Trenton via the Lehigh River through Wilkes-Barre to Lake Erie was 75 miles shorter than the Erie Canal but needed to overcome a greater height. Both of these routes and a third, which would use the Schuylkill River and also go through Wilkes-Barre, were shorter than the one proposed up the Juniata Valley via Pittsburgh to Lake Erie. All three routes would require between 1,700 and 2,900 fewer feet of lockage than the Pittsburgh route. Such a statement raised hopes in Luzerne County that a state canal must logically employ the Susquehanna River.

The Wilkes-Barre meeting agreed that Pennsylvania should not divide its resources and energies in many directions but focus upon one great leading improvement which would benefit the most people and therefore the entire Commonwealth. They asserted that the improvement of the Susquehanna River with a canal making a connection to Seneca Lake would be of greater importance than any other public project. They gave five reasons for taking this position, reasons which had appeared in the local newspaper many times during the previous decade. They recognized the need of Philadelphia to maintain its commercial position and pledged support to the Quaker City. At the same time they said that the city would be nothing without the commerce from the hinterland. The 25 counties in the Susquehanna Valley contained nearly 500,000 people and annually sent a large amount of coal, iron, lumber, manufactured goods, and agricultural produce to the Philadelphia market. With a canal connecting the Quaker City and the

Susquehanna Valley, both the city and the hinterland would benefit from increased economic activity.

The desire of Luzerne County for a lakes-to-sea system following the North Branch of the Susquehanna ran counter to projects emanating from other sections of Pennsylvania. The day before the Wilkes-Barre meeting, the local newspaper printed two notices of other public meetings which supported internal improvements. A Carlisle meeting called for an east-west canal across the Commonwealth and urged other towns and villages to hold similar meetings. From Pittsburgh came news of a town meeting and petitions being circulated in western Pennsylvania counties supporting an east-west canal along the Juniata River to a tributary of the Allegheny River.[16] People in every corner of the Commonwealth envisioned Pennsylvania's grand canal being located along the banks of a nearby stream: one of the branches of the Delaware, Susquehanna, or Ohio river.

Such activity and excitement had been the objective of the Pennsylvania Society for the Promotion of Internal Improvements in the Commonwealth. Responding to the threat of the Erie Canal as well as the promise of prosperity which its initial success brought, 48 of Philadelphia's leading citizens had formed the society in 1824 to gather information concerning transportation projects and technology as well as to bring pressure on the state legislature to undertake a trans-state canal to benefit their city. On 26 January 1825 a Philadelphia public meeting called by the Pennsylvania Improvement Society went on record in support of a water communication from the Susquehanna to the Allegheny River and Lake Erie.[17] Their route would follow either the West Branch of the Susquehanna or the Juniata River.

Announcement of Philadelphia's plan to connect the Susquehanna with the Allegheny River immediately alienated support for the Pennsylvania canal in ten southern counties from York to Greene. In March the state legislature received three petitions from Westmoreland County and a fourth from Franklin County against the state canal recommended by Philadelphia.[18] This region wanted a canal to go from Harrisburg southwest down the Cumberland Valley, across the Allegheny Mountains to the headwater of the Monongahela River, and then to Pittsburgh. The petitions also claimed that an easy connection could be made with the proposed Chesapeake and Ohio Canal.

The ten southwestern counties formed a solid political bloc threatening to destroy Philadelphia's desire for a state-owned and operated transportation system going from the Delaware to the Susquehanna and Allegheny rivers and ending at Lake Erie. In the Pennsylvania House of Representatives these ten counties had 23 votes as compared with the 48 which Philadelphia could count on, votes by representatives whose own immediate self-interest would be served by the canal going through their counties. Philadelphia needed to pick up at least three more votes in the house and preferably more from counties which were willing to sacrifice their own self-interest for the general welfare of the state and city. In a similar fashion Philadelphia needed to get three additional votes in the state senate outside of the path of the proposed canal. To acquire these votes without giving special concessions was just about politically impossible in Pennsylvania due to great sectional diversity and factionalism.[19]

Complicating Philadelphia's problem was the report made by two of the three canal commissioners to the Pennsylvania legislature on 2 February 1825. The Centre County member refused to sign the report of his two Philadelphia colleagues but later submitted a separate statement supporting their position except for the route to be taken. The two Philadelphians favored the shortest route from the Susquehanna to Pittsburgh along the Juniata River even though the mountain barrier to be crossed was higher, whereas the Centre County commissioner liked the West Branch, which did not need as much lockage. He may have also been swayed by the fact that the West Branch route would have benefited his home county more than the southern route.

The 1825 canal commissioners' report proved politically embarrassing for Philadelphia. The Quaker City wanted a unanimous recommendation that the canal be started quickly. The report also caused others to question the judgment of the commissioners. Luzerne County representative George M. Hollenback sent Judge David Scott the commissioners' report. On 26 February Scott replied that he had read the report and found it impossible "to suppress feelings of mortification at the almost total ignorance and stupidity in the gentlemen commissioners which that report indicates. It will be a fruitful source of mirth to the well informed in our sister states." Scott said he hoped provision would be made by law for the appointment of better qualified commissioners to survey the Susquehanna and every other route being considered. Scott, Jacob Cist, and others in Wilkes-Barre carefully followed the course of legislation affecting transportation through the 1824–1825 Pennsylvania legislative session.[20]

In return for certain considerations, Jacob Cist and other Luzerne County leaders were ready to aid Philadelphia. Knowing the results of Philadelphia's public meeting at the end of January, Cist wrote to Mathew Carey, organizer of the Pennsylvania Society for the Promotion of Internal Improvements, about sentiments in Luzerne County. He mentioned the desire to make the Susquehanna navigable from its mouth north to the New York State line with a connecting link to the Finger Lakes. In light of Philadelphia's desire for a project connecting the Susquehanna with the Allegheny River, Cist certainly realized that the Luzerne County project had little chance in the Pennsylvania legislature. If he cooperated with the Philadelphians, his county could get what he considered equally important.

Cist pledged the support of Luzerne County for Philadelphia's project. On 14 February Richard Peters, Jr., a key member of the Pennsylvania Improvement Society, acknowledged Cist's letter.[21] He said that Cist's zeal in the society's cause and the talents which the Wilkes-Barre businessman used to aid the society were known and respected. Peters expressed great satisfaction in knowing that a liberal spirit existed in Luzerne County in favor of a canal from the Susquehanna to the Allegheny River. Recognizing that Luzerne County would not benefit directly from the contemplated project, he praised Cist for putting the welfare of the Commonwealth and the Quaker City above local views and self-interest.

Richard Peters offered the services of the society to help Cist in any way possible. He said the society had no local objects and no particular interests. The entire Commonwealth was the field in which they were disposed to labor. They desired to have worthwhile projects pointed out to them so that they could lend their support.

What did Jacob Cist want? Certainly a canal along the entire length of the Susquehanna River would have benefited his mercantile business, and this project continued to appear in petitions from Luzerne County to the Pennsylvania legislature. Nevertheless, his business had prospered by using the older means of navigation, roads, and turnpikes. Lack of a canal north of Wilkes-Barre to the New York State line would not jeopardize his preeminent mercantile position in northeastern Pennsylvania.

Wyoming Valley's position in the anthracite trade was being challenged by competition from the Schuylkill, Lehigh, and Lackawanna valleys. Each of these areas had new transportation facilities nearing completion, or construction of such projects was about to start.[22] Cist and his associates were keenly aware of the competition they faced from these areas. Like other entrepreneurs, they also anticipated making high profits once facilities were completed for taking coal to market from spring to fall instead of just during short intervals.

In the Wyoming Valley a canal to haul anthracite down the Susquehanna took priority over a trans-state canal. Cist and his internal improvement committee were willing to support the Philadelphia to Pittsburgh project in return for a branch of the state system connecting Wilkes-Barre with the national market. To justify the additional expense, they argued that their 40-mile-long valley contained extensive mines and quarries of stone coal. They counted on the coal and iron trade to pay for the system and said so in petitions to the state legislature.[23]

Repeatedly throughout the spring and summer of 1825 the Wilkes-Barre newspaper emphasized the importance of the coal trade, several times reprinting material from Philadelphia papers. On 18 March the *United States Gazette* took note of the Susquehanna coal trade. About 100,000 bushels of anthracite had arrived from Luzerne County in Philadelphia. Two weeks later the *Wyoming Herald* cited an article from the Philadelphia *Gazette* which said the 100,000 bushels of coal lay on the landing at Mauch Chunk and from 500,000 to 800,000 bushels awaited transportation at the Summit Hill mine. In addition, 90,000 bushels of anthracite were at the head of the Schuylkill navigation, standing ready for shipment as soon as that project was finished in June. During the coming summer Philadelphia anticipated hundreds of thousands of bushels to be mined and shipped so that by fall the Quaker City would take on the appearance of Birmingham, England.[24]

On 15 April another article in Wilkes-Barre's *Wyoming Herald* told of the bituminous coal trade from Clearfield County. During the previous week a great many arks of bituminous had arrived at Harrisburg from this county on the West Branch of the Susquehanna River. According to the article, "The coal trade of Pennsylvania is increasing, and will be immense." One Philadelphia paper noted that Clearfield County had already sent 2,000 tons of bituminous to market by mid-March.[25] So it went week after week, as state newspapers alerted the public to new happenings in the coal trade.

Another major reason for a coal canal from Wilkes-Barre down the Susquehanna Valley appeared in the private correspondence between Judge David Scott and George M. Hollenback. In February 1825 Scott wrote to Cist's brother-in-law at Harrisburg that he felt it was strange that the state legislature should continue to aid the Wurts brothers in their speculation. He thought it likely that everyone in northeastern Pennsylvania might be tied neck and heels and sold to New York by the Pennsylvania legislature. In a previous letter to Hollenback, Scott had gone into greater detail. He noted that the Wurts brothers had sold their coal land and privileges on the Lackawaxen to the Delaware and Hudson Canal Company for $260,000, giving them a handsome profit. New Yorkers had subscribed $1,500,000 in that company which, if completed, Scott feared, would cut off the Lehigh, Schuylkill, and Susquehanna from supplying New York City and all eastern markets with anthracite. The Carbondale mines and Lackawaxen navigation belonged exclusively to New York, and only four or five Pennsylvanians had received any advantages from the sale.

Scott pointed out that from New York City anthracite could be shipped to northeastern states cheaper than from Baltimore and Philadelphia. These two cities would be all that remained for the Lehigh, Schuylkill, and Susquehanna regions to supply. He added that New York City could also send her goods up the Delaware and Hudson Canal and supply all the lower parts of the Delaware and Susquehanna valleys to the exclusion of Philadelphia and ruin its trade with central Pennsylvania. In Scott's mind the Delaware and Hudson Canal would accomplish the same objective for New York City in getting the trade of northeastern Pennsylvania away from Philadelphia as the Erie Canal was doing with the trade of the Midwest.

Scott said that certain privileges should be granted to the Delaware and Hudson Canal Company by the Pennsylvania legislature. If other projects reached fruition, he felt that the Delaware and Hudson people would abandon their project. They would realize that they could get a cheaper supply of anthracite from the Susquehanna and the Lehigh. Scott wanted the Delaware River improved from Lackawaxen south to Trenton so that hard coal would follow its natural path downstream to market. As a consequence, all parts of Pennsylvania would benefit, and its residents would have a fair share of the coal trade which would remain in the hands of Pennsylvanians instead of being transferred to New York City merchants.

David Scott continued to object to the Wurts bill as purely speculative and of benefit primarily to New Yorkers. On 16 March he wrote Hollenback that he had more serious reasons to oppose their bill. As written, the act would give their company complete control over all the anthracite in the Lackawanna Valley. Not one ton would ever be shipped to market upon the Delaware and Hudson Canal by anyone other than the company. The cost of using the locks on the Lackawaxen River would amount to double the value of anthracite in New York City. If the bill included a provision reducing tolls to $.005 per ton per mile, as on other canals, it would not be as objectionable. At the end of March, Scott told Hollenback he was happy to hear that his provision for reducing tolls had been inserted into the bill and would be applicable for the entire length of the canal. Luzerne County coal merchants could profitably send anthracite across the Delaware and Hudson Canal to New York City.

In addition to David Scott, Lewis S. Coryell, another powerful Pennsylvania politician, took an active interest in the Delaware and Hudson Canal as well as a canal along the Delaware River south of Matamoras. In 1815, Coryell, who was a New Hope lumber merchant and had been one of Cist's coal customers during the War of 1812, purchased two tracts of land on the headwaters of the Lackawaxen River in order to supply his business with lumber and other produce.

One of the tracts was a 260-acre farm on the forks of Dyberry Creek, which became the terminal point of the Delaware and Hudson Canal.[26] Like his counterpart in Luzerne County, Coryell and several other merchants in villages along the Delaware Valley wanted to improve both the Lackawaxen and Delaware by means of slack water navigation during the decade following the War of 1812. He became recognized as a local authority on river improvements and in 1822 served as one of the state commissioners to inspect improvements made by the Lehigh Coal and Navigation Company so that they could collect tolls.[27]

Politically, the New Hope lumber merchant was a Jeffersonian Democrat with considerable influence over Bucks County voters. He had been a friend and political ally of Judge Scott for several years prior to 1823 when they both played instrumental roles in helping swing their respective counties behind the successful bid of J. Andrew Shulze for the governorship of Pennsylvania. They were in a position to receive patronage from the governor-elect and worked for each other in securing positions for men in their respective counties.[28]

Like Scott, Lewis Coryell increasingly realized that the coal trade would serve as the best inducement for improving navigation on the Delaware River. In November 1822 Josiah White and Erskine Hazard wrote to him about improving Delaware River navigation and noted that they could now get down the Lehigh safely during the lowest water with coal arks carrying 30–40 tons. They said the improvement of the Delaware was of considerable interest to their company. The following year's operation in the coal business, they claimed, would turn the minds of capitalists to river improvements. There would be no doubt of their stock being promptly subscribed and the improvement being accomplished quickly. If both the Lehigh and Delaware rivers were made navigable from Mauch Chunk to tidewater, the anthracite trade as well as the trade of the northern field would follow their route to market.

The New Hope merchant received similar news the following year from a friend with whom he had shared the desire to improve the Lackawaxen and upper Delaware rivers in 1818. His friend said that public attention locally was absorbed in the survey of the Delaware and Hudson Canal. Such a canal, he felt, would provide water carriage from inexhaustible beds of stone coal in the Lackawanna Valley to New York City. He noted that "DeWitt Clinton and the most leading and monied characters in the city and state of New York are warmly enlisted with such zeal in anticipation [that they can] make of a bushel of stone coal burn 100 bushels of lime in such kilns as they have in England. . . . We have a man in this place that says he has tended such lime kilns in England and Wales. All the poor lime stone lands on the North River are to be made so rich, some wags say, if they stick down a crow bar it would grow full of 10 penny nails."[29] In the Delaware Valley, as elsewhere, Pennsylvania residents were dreaming of the prosperity which would occur with the advent of coal-carrying canals.

During 1824 people living in the lower Delaware Valley were faced with the issue of whether they wanted navigation on their river improved and controlled by the Lehigh Coal and Navigation Company or by the Commonwealth of Pennsylvania. The New Hope merchant became involved in the dispute. In January Josiah White wrote to Coryell about improving the river for boats carrying 100 tons regardless of whether his company or the state undertook the project. In the meantime, Jonathan Fell, president of the Lehigh Coal and Navigation Company, had visited the Bucks County seat of Doylestown to win converts for his company's plan of undertaking the project.[30]

The New Hope merchant sided with those advocating that the state undertake the project. Two Easton politicians traveled to his home to ascertain the sentiments of people in that town as well as the rest of Bucks County concerning the coal company's plan. They had been very happy to hear that they could count on Coryell's support. They wanted him to go to Doylestown to learn what impression the opposition had made and counteract any influence they had exerted. Claiming that residents along the Delaware unanimously supported their position, the Easton politicians maintained that it was necessary for the friends of a free navigation to act, "or we shall be like the people of Germany, not be able to pass our doors, without being obliged to pay Tax, Toll or Tribute to some incorporated company."

Lewis Coryell supported Easton's position because he wanted navigation improved on the entire length of the Delaware River from Matamoras to tidewater. His influence also swung the vote of a Bucks County legislator in Harrisburg on behalf of the Wurts bill in March 1825. Like David Scott, he expected anthracite from the Lackawanna Valley to descend the entire length of the Delaware once both projects were completed and feared that a private company, such as the Lehigh Coal and Navigation Company, might block the coal trade of its competitors to the north. Petitions to the Pennsylvania legislature in 1824 and 1825 showed that a similar sentiment existed throughout the entire Delaware Valley.[31]

In raising the issue of state-owned and operated coal canals, Cist and Scott knew they could count on support in the Delaware Valley. Certainly these people realized there was very little chance that the state canal from Philadelphia to Lake Erie would go through their valley even though they might desire such a project. A similar situation involving the West Branch of the Susquehanna River appeared possible as indicated in the initial report of the canal commissioners. If Philadelphia should influence the state legislature to opt for the Juniata route, counties bordering the West Branch would be left stranded.

The existence of an abundance of bituminous coal in Clearfield and Tioga counties provided a rationale for constructing a lateral canal up the West Branch. Philadelphia and Wilkes-Barre newspapers took notice of these soft coal deposits just as they paid attention to anthracite in the Wyoming Valley. Like the anthracite trade on the North Branch, the bituminous coal trade of the West Branch would more than pay for the cost of constructing and operating a canal. If the Commonwealth did not build a trans-state canal through their region as they wanted, West Branch politicians could be counted upon to support Wilkes-Barre's demand for coal canals as laterals to the main line in order to serve their narrow self-interest. Conscious of the difference between bituminous coal and anthracite, Luzerne County leaders did not fear additional competition in the soft coal trade from Clearfield and Tioga counties. They needed political support in the state legislature and were willing to exchange their votes for similar support.

The coal canal argument did not escape notice in Philadelphia. In February 1825 Richard Peters asked Jacob Cist to send him any ideas worthy of consideration.[32] Whether Cist complied with his request is not known. But high profits anticipated in the coal trade after the completion of transportation facilities on the Lehigh and Schuylkill rivers had caused considerable speculation in the Quaker City during the previous decade. And the Pennsylvania Improvement Society used the coal trade argument to attract support for its cause.

The opening argument in a 4 March communication of the society to the public concerning the trans-state system focused on mineral resources. The society maintained that Pennsylvania possessed greater mineral treasures than any other state in the Union. "In iron & coal, the grand sources of the extraordinary wealth of Great Britain, this state is supremely blest, to a degree certainly not exceeded by any portion of the globe of equal extent." The statement named six Pennsylvania counties with inexhaustible deposits of iron ore. In addition, 20 counties had enough bituminous coal to supply the consumption of half of the entire world. Also singled out for special mention were those counties such as Luzerne, Northampton, and Schuylkill where the supply of anthracite was limitless. The society pointed out that in England coal and iron afforded the chief means of supporting canals wherever they came near such deposits. Tolls from coal and iron would be twice as much as the revenue from all agricultural products. The society estimated the amount of revenue which could be expected from the iron trade and said that an estimate of tolls from coal baffled all calculations.[33] The abundance of coal and iron in Pennsylvania was a unique advantage for the state, something which New York did not have.

The second reason put forth by the society for constructing the Pennsylvania canal was to regain the western trade. Without opening communication with the western states, Philadelphia had no chance of retaining what little remained of her trade with that section of the country. Claiming that its canal would be the chief route of western trade, the Quaker City hoped to regain what it had lost. To Philadelphians, this reason was far more important than developing the internal resources of the Commonwealth.

At the same time Philadelphia faced a problem of losing trade from within the boundaries of the Commonwealth just as it had lost the trade of the Northwest to the Empire State. Several Phildelphia merchants feared that the trade of northeastern Pennsylvania would be drawn to New York City via the Delaware and Hudson Canal if the Commonwealth did not undertake canals along the Delaware and North Branch of the Susquehanna. They were already competing intensely with Baltimore for the trade of the Susquehanna Valley. While Cist, Hollenback, and other Wyoming Valley merchants traded primarily with Philadelphia, their business could be carried on just as easily with merchants in Baltimore and New York City. Confronted with this situation, Philadelphia's mercantile community was ready to embrace a far larger transportation system than it originally wanted. Philadelphia needed the votes of the northeastern and north central counties in the state legislature in order to get the trans-state canal, and it would do just about anything to retain the commerce of these sections of the state. Anticipated tolls from the coal and iron trade could justify the additional expense.

By the summer of 1825 an unwritten compromise had been reached involving the selection of state canal commissioners and a canal convention to determine the priority of projects. During February and March the state legislature spent much time discussing the canal commissioners' report. Several counties resented the fact that all three of the 1824–1825 commissioners lived along the line of the proposed routes. How could Philadelphia ask other sections of the state to put the welfare of the Commonwealth above their own self-interests when, in fact, they did the opposite. If the commissioners were to serve the welfare of the entire Commonwealth, they should come from areas outside of the proposed routes. Many felt that enlarging the number of commissioners from three to five would give greater representation and provide for a better report than they had received from the first commissioners. The legislature finally passed a bill endorsing a canal going from the Susquehanna to the Allegheny River and enlarging the number of commissioners to five.[34]

Governor J. Andrew Shulze had the power to appoint the five canal commissioners whose advice would be weighed heavily by the next legislature. If the commissioners rendered a unanimous opinion, the legislature would be under pressure to follow their recommendation.

As directors of the Pennsylvania Improvement Society, Mathew Carey and his friends were concerned about whom the governor would appoint. Realizing that current commissioners had influential friends throughout the Commonwealth whom he did not wish to alienate, Carey did not encourage blocking their reappointment. He wanted John Sergeant, president of the Internal Improvement Society, appointed to the commission. With Sergeant in that position, the society could accomplish its objective. He would not find it too difficult working with other commissioners. Carey went to Harrisburg to talk with Governor Shulze about the wishes of the society.

One of Carey's colleagues singled out a politically influential Philadelphia merchant whom he did not want as a canal commissioner. He disapproved of this merchant because he was a director of the Schuylkill Navigation Company and represented the interests of this company as well as the Union Canal Company. Both private corporations opposed the canal through Lancaster County, one of the routes to be surveyed by the commissioners. To get the votes of Chester and Lancaster counties in the state legislature, the Internal Improvement Society advocated transportation facilities crossing these two counties from Philadelphia to Harrisburg.[35]

Carey must have had some luck with the governor. John Sergeant replaced one of the Philadelphia canal commissioners and the Schuylkill Navigation manager remained off the board. Shulze also appointed a party leader from Chester County as a commissioner, thus increasing the chances for support for the project crossing Chester and Lancaster counties. In addition, Governor Shulze appointed Judge David Scott to the commission.

In 1825 all of the canal commissioners lived along the line of the proposed route except David Scott of Luzerne County. In addition to Scott, Sergeant, and the commissioner from Chester County, one commissioner came from Philadelphia and another from Beaver County. None of the previous commissioners were reappointed, nor did any county embracing the Juniata and West Branch valleys have one of its residents on the board.

In commenting on the character of the commissioners in an editorial published on 29 April 1825, the *Pennsylvania Intelligencer* felt that public interest was perfectly secure because the canal commissioners would not be unduly influenced in selecting the principal east-west route. The geographical makeup of the board virtually guaranteed this fact, and Luzerne County's internal improvements committee had pledged its support to the route. But the selection of Scott indicated that more than a single main route was under consideration and would be acceptable.

In corresponding with legislative delegate George M. Hollenback, Scott had set forth his views. In February 1825 Scott wanted the entire Susquehanna River surveyed for a possible canal. He hoped that the law would provide for commissioners to survey the Susquehanna and every other route under consideration. After receiving the bill for appointing commissioners as passed by the Senate, he told Hollenback that it was certainly better for northeastern Pennsylvania than the one originally reported. Scott wanted further modification in the bill so that the commissioners would be left entirely free to examine canal routes in any part of the Commonwealth. Such a provision might be made without endangering the bill. One week later he repeated his point that the bill give the commissioners some discretionary powers over where they would make their surveys.[36] The bill, as finally enacted, provided for such discretionary power.

Fear that the Commonwealth would undertake too many projects was evident among leaders of the Pennsylvania Society for Internal Improvements. On 9 May 1825 John Sergeant wrote to Mathew Carey to be cautious with the information he had collected. While all of it was valuable and worthy of consideration, they had reached a point where it became necessary to simplify their views as much as possible before presenting their plans to the public. Sergeant was concerned that the public mind might be distracted by the diversity of plans and arguments related to local schemes injurious to the main cause. He would be happy to meet with the delegates to the canal convention as soon as possible but not until after he had met with the other canal commissioners.

The canal commissioners set the tone for the forthcoming canal convention. When they convened on 4 July they elected Sergeant chairman and agreed on the priority of the east-west route. It is not known whether Scott raised the issue of coal canals at the commissioners' meeting. More than likely he expressed the same ideas at this meeting which had circulated in Luzerne County during the previous decade as well as what he had told George Hollenback in private. As president of the Pennsylvania Society for Internal Improvements, Sergeant knew about these projects, and other members of the commission may have been equally aware of them. While Sergeant wanted to simplify plans as much as possible, arguments for coal canals would be heard at the Harrisburg convention.

In August 1825 Luzerne County sent its foremost anthracite expert to the Harrisburg canal convention. Jacob Cist's reputation as an authority on this subject was clearly recognized at the meeting.[37] Others attending the convention had also been involved in the development of Pennsylvania's coal trade for more than a decade. In addition to Cist, two other delegates had marketed anthracite in Philadelphia during the War of 1812. James Ronaldson, the typefounder who had been one of Cist's steady anthracite customers since the war, was one of Philadelphia County's seven delegates. While none of the counties in the Lehigh Valley sent representatives who were managers or large stockholders of the Lehigh Coal and Navigation Company, the city of Philadelphia as well as Berks, Montgomery, and Schuylkill counties had a total of 8 of their 16 elected delegates clearly identified with the Schuylkill coal trade. Two of them had been directors of the Schuylkill Navigation Company since its creation in 1815. At least 70 of the 117 delegates from the 48 counties represented at the canal convention had some knowledge of coal. Many of them had already invested sizable sums in the trade.[38] When Jacob Cist spoke to the convention about coal, most of the audience had previously given the subject serious consideration and shared his views about the potential of the coal trade.

On 4 August the canal convention met in Harrisburg. John Sergeant, head of the Philadelphia city delegation, made the opening speech, emphasizing that Pennsylvania had always been among the leaders of the internal improvements movement even though New York had received most of the attention. While the Empire State had constructed the Erie Canal, Philadelphia capital had gone into the Schuylkill Navigation Company and the Lehigh Coal and Navigation Company. Both projects had greatly reduced the cost of transportation in their respective regions. The time had arrived to undertake a canal linking Philadelphia with the Ohio River and the Great Lakes in order to regain the western trade. Sergeant noted the high mountains separating the East and West. These mountains, he pointed out, contained the essential resources of coal and iron for overcoming all obstacles. Sergeant recommended building the main line first and constructing lateral branches later. He moved that the convention select a committee to draft resolutions for general discussion.[39]

In keeping with Sergeant's suggestions, the next day a committee presented four resolutions: first, that transportation facilities be constructed from the Susquehanna to the Allegheny and Ohio rivers, and then to Lake Erie; second, that the state undertake the project; third, that local projects be put aside until the main objective had been achieved; and last, that sister states be encouraged to construct canals from the Potomac to the Ohio and from the Finger Lakes to the North

Branch. James Buchanan, a Lancaster County delegate, offered amendments to the original resolutions. He wanted the fourth point substituted for the second and third and added another resolution that the convention approve the act creating canal commissioners. They should select the best route between the East and West.[40]

Buchanan's proposals gave rise to considerable discussion which lasted the entire meeting. Opposition to the original resolutions came from Lancaster County, the southern counties, and Easton. While the motives of the southern counties can easily be attributed to sectional demands, Buchanan's action is more difficult to explain. In playing down the importance of the east-west canal, he helped open the question for a more comprehensive state system of a main line and branch canals. The canal commissioners were empowered to investigate more than the main line, and Chester and Lancaster counties wanted the route of a railroad across their counties surveyed. To win support of the Schuylkill Valley delegates, in his opening speech Sergeant had mentioned using the route of the Schuylkill and Union canals to reach the Susquehanna. This proposal may have also alienated the Lancaster County delegates.

The other main center of opposition came from the Delaware Valley counties of Lehigh, Northampton, Pike, and Wayne. An Easton delegate voiced his opinion that a state-owned and operated canal would raise taxes too much.[41] He represented conservative Pennsylvania German constituents who had a strong tradition of opposing taxation. If he had hoped to win support for his position from other delegates representing rural conservative agrarian interests who disliked increased taxes, he failed. Pennsylvania's farm population realized that better transportation facilities would raise real estate values and bring them a greater return on the sale of farm land.

These delegates from eastern counties may have also opposed the resolutions because they wanted a more comprehensive system than just the main line. The Easton delegate may have revealed his true position when he advocated improving navigation on the state's principal streams, feeling that the anthracite trade down the Delaware River would pay the bill for his branch of the system.[42] If this occurred, the taxes of his constituents would not be raised, and they would have better transportation facilities to take agricultural produce to market.

One potential voice of opposition was not heard and consequently did not receive the wrath of Philadelphia newspapers and Mathew Carey.[43] Jacob Cist and other delegates from north central Pennsylvania went along with Philadelphia. Defection of 18 delegates from 11 counties in north central Pennsylvania would have cut the voting margin for a state system from 75 to 60 percent and greatly reduced the chances of the legislature providing for the project. Assuming that this group followed Philadelphia blindly in anticipation that the east-west canal would come their way fails to reckon with the facts. Many of these delegates realized that the main line would not come near their locality. Yet they voted overwhelmingly in support of the state system, making it appear that the Harrisburg canal convention stood solidly in support of the Pennsylvania canal.

Luzerne County's position provides a reasonable explanation of why this 11-county bloc backed Philadelphia. At the convention Jacob Cist talked about the abundance of anthracite in the Wyoming Valley. His careful examination of the coal formation and potentially profitable mines permitted him to speak persuasively to a sympathetic audience about a canal for the North Branch of the Susquehanna River. His claim that the anthracite trade would more than pay the cost of constructing and operating this artery did not fall on deaf ears.[44] Others in the room entertained similar ideas about the coal deposits of their respective regions.

While advocating a more comprehensive system than just the main line, Cist was willing to vote for Sergeant's resolutions. He fulfilled a pledge made the previous January and, consequently, was in a position to forge a compromise. He wanted a state transportation system since private capital in his county could not afford to undertake his project and had nothing to lose in supporting Philadelphia. Furthermore, the Harrisburg canal convention did not have the final say about the extent of the system. That decision rested with the canal commissioners and the state legislature. In both bodies Luzerne County exercised considerable influence and ultimately achieved its goal.

Three years passed before the final details of Pennsylvania's transportation system were finalized. The main responsibility for the system rested with the canal commissioners. David Scott served on the board from 1825 to 1829 and during his last year was its chairman. He played a major role in creating a comprehensive system. In the 1825–1826 session their attention focused on the link between the Susquehanna and Allegheny rivers, and the board recommended the Juniata route. The North Branch did not come under consideration by the commissioners until 1826.

On 10 March 1826 the commission agreed that Scott would assume responsibility for surveying the North Branch. Two months later he employed a Luzerne County surveyor and examined the route from Northumberland to the New York State line. Accompanying the surveyor's report were Scott's reasons for undertaking the project. He repeated arguments which had been expressed in Luzerne County meetings for the previous eight years. All the reasons for improving river navigation received his attention. His most important point dealt with the anthracite trade. Coal seams from 5 to 40 feet thick covered an area from the head of the Lackawanna Valley to Shickshinny. These coal measures would not be exhausted in 10,000 years. He estimated the export of anthracite down the Susquehanna at 50,000 tons annually and predicted it might go as high as 100,000 tons a year. At a rate of 1 percent per ton per mile, the anthracite trade would produce an annual toll of $59,000, more than enough to pay the construction debt.[45]

In 1827 the canal commissioners accepted Scott's recommendations. They visited Wilkes-Barre and upon seeing the thick coal seams in the Wyoming Valley agreed that the Wyoming Valley anthracite trade would pay for the lateral branch. The 161-mile section from Northumberland to the New York State line would require 377 feet of lockage. They estimated the cost to exceed $1,000,000 and recommended that the Commonwealth immediately undertake the section south of Wilkes-Barre. By 1827 the canal commissioners were convinced that the coal trade of interior Pennsylvania would pay for the entire system. This article of faith, first expressed by Jacob Cist at Luzerne County meetings and pushed by David Scott, became part of the official canal commission reports sent annually to the state legislature.

Luzerne County gave the canal commissioners full support in their petitions to the Pennsylvania legislature. Every petition endorsed the general system of internal improvements. They pledged support for other lateral branches, anticipating cooperation in getting funds for the North Branch Canal. The coal trade was always used as the chief reason for building the canals, and the petitions maintained that the coal trade would pay the costly construction bill.[46]

Luzerne County received support for its demands in the Pennsylvania legislature. In March 1827 the chairman of the House Committee on Inland Navigation reported a bill favoring a comprehensive system, namely the east-west canal following the Juniata Valley, the North Branch Canal to the New York State line, a lateral along the west branch to the Sinemahoning Creek, the Erie Extension Canal, and the Delaware Division Canal. After lengthy debate in the House of Representatives reflecting the sectional divisions observed at the Harrisburg canal convention, the bill passed the house by a vote of 53 to 45. Counties in the Schuylkill Valley opposed the bill because they did not want additional competition in the coal trade from the northern field, and Lancaster and Chester counties voted against the bill because it did not provide for a railroad from Philadelphia to Columbia. The voting pattern in the Senate showed the same split, with 19 for the bill and 15 against it. An even greater margin of support for the system was picked up the following year when the canal commissioners recommended a railroad crossing Chester and Lancaster counties. Two-thirds of the legislators in the Senate and House of Representatives supported the package which ultimately became the public works.[47]

On 2 February 1828 the chairman of the House Committee on Inland Navigation expressed the general sentiments regarding the state transportation system. He mentioned the benefits which would come from the east-west facility. In addition, he paid particular attention to the potential coal trade. The Schuylkill mines were not yet in full production, he noted, and had produced tolls of $60,000 during the previous year. He said:

> The committee cannot avoid adverting especially to the resources Pennsylvania has in coal, the most valuable of all articles as tonnage for canals or railroads. The engineers of the Lehigh Coal and Navigation Company have made a calculation to show that the coal trade, when the population within ten miles of tide are supplied, will pay to Pennsylvania four million dollars annually in the shape of tolls on the improvement. It is said there are annually brought into use in Great Britain twenty millions of tons of coal, and the consumption of London alone exceeds one million tons. The West Indies and perhaps France will hereafter consume Pennsylvania coal. The market of the United States is open from Maine to New Orleans, and at this time Pennsylvania coal is carried by inland navigation more than 100 miles to Philadelphia, and thence carried by sea to Richmond and on account of its superior quality, consumed there, within a few miles of the Virginia coal mines.[48]

The ideas of Jacob Cist and others had become official state policy as they were voiced in the legislative chambers and enacted into law. Accepting the lateral branches in order to develop its coal deposits meant that Pennsylvania enlarged the original commitment which Philadelphia had made after the War of 1812 to open the southern anthracite field. The Wyoming Valley acquired a canal to handle its coal trade.

The North Branch Canal played a vital role in making Cist's dream a reality. Matthias Hollenback helped break ground for the lateral on 4 July 1828, and six years later it opened for trade. Nearly a decade passed, however, before coal producers shipped more anthracite on the canal than on the Susquehanna River.

The anthracite trade on the North Branch Canal greatly exceeded the original estimate, and one would expect that the coal trade paid for the cost of constructing and operating the canal. But unforeseen difficulties raised the cost of construction far above Scott's original estimate of $1,288,000 for the lateral from Northumberland to the New York

State line. The section south of Wilkes-Barre cost $1,623,117, and the Commonwealth spent more than $4,500,000 on the section above Wilkes-Barre before selling the public works. The anthracite trade did pay the operating expenses for the canal. From 1830 until Pennsylvania sold the canal in 1858, the Commonwealth spent just over $1,000,000 to operate the lateral and took in $2,301,979 in tolls with most of the revenue coming from the anthracite trade.[49] The canal remained in operation for nearly 50 years after the state sold it to private enterprise, and its use for hauling anthracite continued until the turn of the twentieth century. The North Branch Canal achieved its primary objective as envisioned by Cist. It served as a vital link connecting the Wyoming Valley coal deposits with the national market and stimulated the development of the anthracite trade.

8 Cist's Ultimate Dream: An Anthracite Railroad to New York City

While Jacob Cist and his associates recognized that the easiest route for shipping Wyoming Valley anthracite to market was by means of a canal down the Susquehanna River, they considered a transportation system which would cross the 300-foot-high mountain that served as a barrier for taking hard coal directly southeast to New York City. Sometime between 1815 and 1825 Cist made a table showing the heights and distances involved in three canal routes from the Luzerne coal formation to New York, the first using the Delaware and Hudson Canal and two others linking tributaries of the Susquehanna with the Lehigh River. By using the latest British stationary steam engines and tramways to cross the mountain, several persons told him that he was at least 50 years ahead of his time.[1]

A friend sent Cist's notes for these canals to one of New Jersey's leading canal advocates, George MacCulloch of Morristown, promoter of the Morris Canal. On 6 February 1825 MacCulloch wrote to Cist that he had received a copy of his plans for the several routes to connect the Wyoming coal formation with New York City. While MacCulloch admitted he knew little about the terrain of northeastern Pennsylvania, he said that he knew the topography of New Jersey. He told Cist that the shortest way to New York was through the Morris Canal. The distance from Easton to New York Bay was 85 miles over an elevation of only 850 feet. MacCulloch felt that Cist would do best to concentrate his efforts upon a passage from Wilkes-Barre to Easton for a connection with the Morris Canal. He hoped that the Morris Canal would be completed within four years.[2]

The year 1825 was a pivotal year in transportation history. As noted previously, the instant success of the Erie Canal, which at that time was the longest canal in the world, caused a strong reaction in Pennsylvania, and Jacob Cist numbered among those supporting canals for the Commonwealth's public works. At the same time he was keenly aware of developments in England concerning railroads, particularly the anticipated completion of the Stockton and Darlington Railroad, which would carry both coal and passengers.[3]

Cist was not the only Pennsylvanian weighing the merits of canals and railroads as the best means of transportation. In March the Pennsylvania Society for Internal Improvements sent a young engineer to England to investigate the technical details of canals, railroads, coal mines, and other scientific inventions. Meanwhile, other Philadelphians accumulated information about transportation, anticipating means that should be used to connect their port with Pittsburgh.[4]

On 1 August 1825 a pamphlet appeared in Philadelphia advocating the use of railroads rather than canals in Pennsylvania. It said that on a level railroad, one horse could pull 10.5 tons at the rate of four miles an hour. By 1815 England and Wales had several railways drawing coal cars over flat cast iron rails. Using high pressure steam engines, British engineers had begun to experiment with small eight-ton locomotives of eight horsepower for drawing 45 tons of goods at eight miles an hour. These locomotives would pull trains over grades as well as on the level. One such railroad was planned to go 400 miles from London to Edinburgh, and another entrepreneur was considering carrying coal by land from Newcastle to London by means of a railroad. The author of the pamphlet said that a speculative mood currently gripped the British Isles for using railroads with locomotives. He believed that the time had also arrived for their construction and use in the United States.[5]

Between February 1825 and the following September, George MacCulloch had changed his mind about the type of transportation to connect the Luzerne coal formation with New York City. On 20 September 1825 MacCulloch wrote to Cist that he had become disillusioned with the gentlemen who controlled the Morris Canal. He said they could not command the supply of coal to the New York market and that they did not go through the most populous district of northern New Jersey. Because they had abandoned his plans, MacCulloch for some months had contemplated a railroad from the anthracite formation to New York City passing in the vicinity of Morristown. He felt that such a railroad would undersell every other form of competition. If a liberal charter could be procured in Pennsylvania, he said that New Jersey would not put any obstacle in the way. A railroad corporation would readily be subscribed in New York City, especially if Cist and others in the Wyoming Valley gave them strong support.[6]

According to MacCulloch, the Mauch Chunk and Wilkes-Barre areas seemed to have the only great coal quarries to be taken into consideration as the beginning points for a railroad. He thought the Lehigh Coal and Navigation Company would decline cooperating in any project which might undermine its plans. MacCulloch said that the Lehigh company currently demanded a price for coal which would never be paid in New York City. For these reasons MacCulloch said it was apparent that Wilkes-Barre was the most eligible beginning point for an easy and straight line to New York City.

MacCulloch suggested that Cist make plans for a railroad from the Wyoming Valley to the Delaware Water Gap. From the Delaware Gap an easy route could be constructed across New Jersey near the vicinity of Hackettstown and Morristown to the New York Bay. MacCulloch told Cist he only needed to explore the various routes through Pennsylvania to Columbia, New Jersey. He would undertake the survey in northern New Jersey. Both MacCulloch and Cist realized that the principal commodity of the railroad would be anthracite. MacCulloch estimated that about 115,000 tons of hard coal would annually come across this route. Adding a considerable increase of the iron production in northern New Jersey to this trade, MacCulloch believed that the project would prove extremely profitable. He thought that the railroad might also be extended westward to Erie on the one hand or Pittsburgh on the other. But he felt it would be safe to construct just the line from Wilkes-Barre to New York City in the beginning, leaving the other lines for later.[7]

MacCulloch wanted to obtain a charter from New Jersey for this company. He needed three names for directors from Pennsylvania as well as three from New Jersey and three from New York. Since the New Jersey legislature met in October and MacCulloch needed to have the bill ready by 1 November, he wanted Cist to act immediately. During the coming months they would have to obtain charters from Pennsylvania and New Jersey. In the spring they could make the surveys and estimates so that a company could be formed by the end of 1826.

Cist complied with MacCulloch's request. On 20 October MacCulloch said that he had received a recent letter from Cist and its accompanying document—the charter which would be presented to the New Jersey legislature. MacCulloch told Cist that a friend would bring a sketch of the charter to be proposed to the Pennsylvania legislature which was modeled on that of the Morris Canal. The charter would contain everything they wanted, MacCulloch noted, as the whole operation would be jeopardized unless Pennsylvania and New Jersey laws coincided. MacCulloch felt that Cist had better not present his bill until they knew what shape the New Jersey law would take. He told Cist he would send a copy of the law as soon as possible. It would help facilitate the passage of a similar law in Pennsylvania. To assist his plans, Mac-

Culloch wanted to receive from Cist a letter stating the magnitude of the law to be attained by the scheme and its bearing not only upon the northern part of the Union, but also upon the states of New Jersey, New York, and Pennsylvania. The practicality of a route from Pittsburgh to the Delaware Water Gap should be mentioned as well as the willingness of Pennsylvania to accede to the plan.[8]

MacCulloch wanted to capitalize the corporation at $1 million with a clause to increase the capitalization to $2 million if necessary. His reason for inserting so large an amount was that the capital of either of their bills might be needed to construct the whole line from Pittsburgh to New York City. MacCulloch said that it was indispensable that the entire project should fall into the hands of the same body of stockholders. To get subscribers, a majority of the stock had to be acquired from New York City capitalists. MacCulloch needed the names of commissioners to organize the company. Three would be from Pennsylvania, three from New Jersey, and nine from New York. If this arrangement suited Cist, MacCulloch wanted him to send the three Pennsylvania names immediately. He would then go to New York City to try to enlist nine leading merchants. He felt he already had the cooperation of two or three financiers to back the project, including a Luzerne County landowner to whom Cist had sent samples of anthracite during the War of 1812. MacCulloch wanted Cist to serve as one of the commissioners from Pennsylvania, and he would be one from New Jersey.

On 28 October Cist replied to MacCulloch, proposing a company which would own coal land in the Wyoming Valley and have a railroad running from the mines to New York City. The Wyoming Valley projectors of the company, with a view toward establishing a respectable and efficient coal company, had selected a number of very valuable mines. Their selection was made with particular reference to the advantages of working and draining the mines as well as in regard to their proximity to the Susquehanna River and the line proposed to New York City. They were willing to offer their property for sale to the company. Like the Wurts brothers, the object of disposing of their property was to enable them to acquire the essential capital for working effectively and transporting the coal to market. They wanted to reserve the right to use this money to purchase stock in the new company as might be agreed upon between themselves and the other stockholders.[9]

As noted in the earlier chapter on the Wyoming Valley, Luzerne County merchants lacked the funds to work their mines on a large scale. The leading hard coal merchants had agreed to form an association to exploit their resource. They hoped to trade the lands and mines they had acquired during previous decades for capital to construct the costly means of transportation from their valley to a major coal-consuming market. They wanted to make money by selling their mines as well as from the profits this anthracite corporation would produce. The backers of the company included John Smith, who agreed to put his entire coal bed into the project. In addition, Cist's large mine south of Wilkes-Barre would be sold to the coal corporation. Three other Wyoming Valley merchants were also willing to join with Smith and Cist. Two of these three merchants offered their land and ark yards, and a leading Plymouth merchant agreed to back the project by selling a good coal bed worth $4,000.[10]

Jacob Cist went one step further. He constructed a map and sent it to MacCulloch showing the principal landholders in Luzerne County who owned tracts south of the anthracite formation. The map gave the location of land owned by Jacob Cist and Matthias Hollenback. In addition, seven other important landowners rated special notation by Cist. They included the estates of three leading Philadelphia financiers, tracts owned by the Bank of North America, Hope and Company of Amsterdam, an English member of Parliament, and a New York financier.[11] If a railroad crossed this terrain, the value of the land would increase. Some of the men and financial institutions could be counted upon to subscribe stock in the corporation.

Within a week MacCulloch wrote to Cist about matters which the Wilkes-Barre merchant had proposed in his correspondence. MacCulloch found Cist's map interesting. He said that Cist's name as well as the two other Luzerne County gentlemen would be introduced into the New Jersey bill which had already been drawn up. Cist wanted the bill presented first in Pennsylvania. While MacCulloch understood his reasoning, he felt it best to push ahead in New Jersey since the time was appropriate. If they gave others an opportunity to organize resistance on behalf of the Morris Canal Company, the difficulty of obtaining a bill from New Jersey might prove insurmountable. Therefore, MacCulloch said he would go ahead, taking care not to hurt any proposed operation in Pennsylvania unless absolutely necessary.

MacCulloch told Cist that he would start the next day securing support among his New York City friends. If he could acquire the support of five or six capitalists to become commissioners along with those from Pennsylvania and New Jersey, he would go to Trenton within the next ten days. MacCulloch promised to keep Cist constantly informed about the project. Should MacCulloch's New York friends be unwilling to give support, he would wait and let the process begin in Pennsylvania as Cist desired.

Evidently MacCulloch succeeded in getting his bill from New Jersey, and then Cist submitted a bill to the Pennsylvania legislature. On 16 December 1825 Jacob had his wife write to George M. Hollenback in Harrisburg. Cist said that only the bill sent to his brother-in-law should be reported because it contained the provisions necessary in order to get the backing of the gentlemen who proposed purchasing the stock. If the bill was reported without any changes, the project would succeed. On the other hand, if those sent by two other politicians were enacted, the entire plan would be destroyed.[12]

Sarah Cist told her brother that Jacob was too ill to write. She said that the business of the railroad currently occupied an overwhelming amount of his time and energy. She though that his constant anxiety on the subject only increased his sickness, which appeared to be a mixture of gout and an infection of the liver. His health had declined steadily since George had gone to Harrisburg. His spirit and eyesight were greatly impaired. Within a week George Hollenback headed home where he found Jacob extremely ill but not confined to his bed. However, Jacob Cist's health took a turn for the worse. On 27 December he was confined to his bed, and three days later he died.[13]

The death of Jacob Cist on 30 December 1825 caused a setback to the development of the anthracite trade in the Wyoming Valley. Entrepreneurs in the Schuylkill and Lehigh regions had their projects well under way. George Hollenback and others inherited Jacob Cist's dreams as well as his numerous notes accumulated during the previous decade. Without Cist's vision and prestige, however, they lost much of the impetus for their coal and railroad company.

On 19 January 1826 George Hollenback received a letter from MacCulloch in which he lamented the death of Cist but hoped that Hollenback would carry on the work of getting a railroad from Wilkes-Barre to New York City. MacCulloch wanted to know the fate of the project and asked Hollenback to send him a copy of the bill if it was passed by the Pennsylvania legislature. With this information he would carry on the business of gaining support from New York financiers. Dr. David Hoosak, one of New York's leading financiers, was still interested in the project. In 1807 he had purchased 8,000 acres in Luzerne County from Matthias Hollenback. The land which he owned was south of the coal formation along the route of the railroad. As one of New York City's principal financiers, he was interested in constructing the railroad and forming the coal company.

The death of Cist, nevertheless, altered plans for a railroad from Wilkes-Barre to New York City. George Hollenback was willing to carry on the project and received another letter from George MacCulloch. Henry M. Drinker, a Philadelphia financier and one of the landowners mentioned on Cist's map, also wrote to Hollenback. He too lamented the death of Cist. On 5 January 1826 he commented to Hollenback about the plans for a railroad from the Wyoming Valley to Delaware

Water Gap. Two weeks later a resident of Stroudsburg, Pennsylvania, wrote to George Hollenback suggesting that Drinker replace Cist as one of Pennsylvania's commissioners for the projected railroad.[14]

Henry Drinker replaced Cist as one of Pennsylvania's commissioners, and with members of the Scranton family he spearheaded the movement for a railroad from the Susquehanna River to the Delaware Water Gap. Over 25 years passed until the Delaware, Lackawanna, and Western Railroad finally completed the line first proposed by Cist, and it took two more decades until that railroad accumulated 25,000 acres of coal land to make it one of the great anthracite corporations.[15] The railroad went from Scranton instead of Wilkes-Barre to New York City. The merchants of the Wyoming Valley had certainly been accurate in forecasting that it would be at least 50 years until Cist's dreams were fulfilled.

Meanwhile George M. Hollenback remained one of Wilkes-Barre's leading coal merchants until he died in 1866. To convey his anthracite to market, Hollenback relied on the North Branch Canal as well as a railway from Wilkes-Barre to the head of Lehigh navigation at White Haven. This railway, using stationary steam engines to cross the 300-foot-high barrier south of Wilkes-Barre, followed one of the routes projected by Jacob Cist to get Wyoming Valley coal to market.

Conclusion

Jacob Cist was typical of most early nineteenth-century merchant entrepreneurs. Like a majority of these businessmen as well as other pioneers in the anthracite trade, he came from a mercantile background and had been trained in the ways of business. Cist's business career spans a time period when the national economy was in a state of change, and he seized upon opportunities in his immediate locality to become an innovator. In this respect, too, he was similar to other entrepreneurs of his day.[1]

As a partner in the leading mercantile firm of northeastern Pennsylvania, Jacob Cist was keenly aware of the marketing process, and he understood the present and future needs of the Keystone State as well as Philadelphia. From his father and uncle he acquired the vision that Pennsylvania anthracite would become one of the major fuels for Philadelphia and other Atlantic coastal cities once a market had been established. Circumstances contributed to Cist's success in creating a market for Pennsylvania hard coal. The War of 1812 caused a fuel crisis in Philadelphia when a severe shortage of coal forced its price to increase about 300 percent. The extremely high price provided incentive for Cist to initiate the hard coal trade. In subsequent years he never lost sight of the profits to be made once the anthracite trade had been established on a permanent basis, and he candidly admitted this primary motive to close friends on a number of occasions.

While potential profits served as Cist's major motive for initiating the anthracite trade, other factors must be taken into consideration. Several historians have noted cultural factors as the motivating forces of nineteenth-century entrepreneurs. Cist lived at the time of the American enlightenment when belief in material progress pervaded society.[2] He shared this common belief with his contemporaries and felt that the anthracite trade would develop almost automatically if certain individuals led the way. Social status has also been cited by historians as a motivating factor for entrepreneurial activity, and Cist's behavior falls into this category. According to this theory, merchants occupied elite positions in society to affect economic change.[3] Cist's social and economic position in the Wyoming Valley as well as the rest of Pennsylvania placed him in a position of leadership. No doubt, he used his social status to achieve his goal.

To achieve his objective, Cist had to encourage four distinct groups to take an interest in Pennsylvania anthracite. The first was the consumer. Philadelphia craftsmen had burned both hard and soft coal throughout the colonial period with the variety coming from Kilkenny, Ireland, closely resembling Pennsylvania anthracite. When the fuel crisis occurred in 1813 and 1814, Cist made plans to visit every type of craftsmen in the Quaker City who used either Virginia or foreign coal for fuel and talk with them about experimenting with his hard coal. He succeeded in persuading a wide variety of craftsmen to use anthracite. Once an abundant supply of Pennsylvania hard coal arrived in their city at an inexpensive price, it would become the primary fuel among urban craftsmen. Cist also paid attention to the large market for his coal among Philadelphia's residents.

Philadelphia's coal merchants formed the second group with which Cist had to work to establish a market for his anthracite. Like the urban craftsmen, they knew about the wide variety of coal and had sold it in increasing amounts since the American Revolution. During the fuel crisis of 1813–1814 Cist convinced at least two coal merchants to handle his anthracite, and they were willing to continue selling it after the War of 1812. He had worked with several of the city's leading coal merchants who wanted to sell his product, thereby helping to create an important link in the marketing process.

The scientific community was the third group with which Cist worked to create a demand for Pennsylvania anthracite. Businessmen relied upon expert advice before investing in mining and costly transportation projects. Although Cist was primarily a merchant, he also laid claim to being an amateur scientist. American scientists such as Benjamin Silliman, James Renwick, and Thomas Say recognized Cist as the foremost expert on the Pennsylvania anthracite formation and in private correspondence told him so. Cist more than anyone else brought the attention of the scientific and business communities to the abundance of anthracite in northeastern Pennsylvania. Calculating that the entire Pennsylvania anthracite formation contained from 8 to 10 billion tons, Cist said there was sufficient fuel in this region to take care of the economic and urban needs of the United States once the Commonwealth of Pennsylvania or Philadelphia capitalists constructed the costly transportation projects to get this fuel to market.

The financiers formed the final group necessary for establishing a market for Pennsylvania anthracite because they would put up the money for costly transportation projects. Like leading American scientists, Philadelphia financiers and Pennsylvania politicians respected Cist's knowledge of the anthracite formation. He helped establish two transportation projects and died while planning a third.

Cist realized that the anthracite trade would become the leading sector for the economic growth of northeastern Pennsylvania although he lived during the time span which one economist has called the preconditions for economic take-off.[4] Aware of the impact coal had on England, he believed that Pennsylvania hard coal would provide a similar stimulus on his region. His dreams came true as the annual output of anthracite increased just about every year in the nineteenth and early twentieth centuries until it reached a peak of 100 million tons in 1917, giving employment to more than 150,000 hard coal miners. From 1870 to 1920 more than 50 percent of the hard coal produced came from the northern field.[5]

Due to coal mining, from 1870 to 1920 Luzerne ranked as the third most populous county in the Commonwealth, while Lackawanna County, erected in 1878, attained fourth place by 1900. Two of Pennsylvania's top-ten cities—Scranton and Wilkes-Barre—grew up in the northern field, each becoming a standard metropolitan area by 1900.[6] While mining employed the greatest percentage of their population, these cities were also major commercial centers and had diversified economies. Their key industries supported coal production, and in Scranton and Wilkes-Barre iron production consumed a large amount of anthracite. Certainly this achievement more than fulfilled the expectation of Cist.

Cist also foresaw the importance of Pennsylvania anthracite to the urban industrial growth of the United States. Like many of his contemporaries, he noted the need to save time and labor and explained how anthracite would help. However, his death in 1825 came too soon for him to realize that Pennsylvania anthracite would cause the beginning of the Industrial Revolution in the United States.[7] Nor would he realize that at the end of the nineteenth century Pennsylvania anthracite would be shipped by boat and rail to just about every major city in the country and become the most important domestic fuel for urbanizing America.[8]

Jacob Cist lived during the age of American mercantilism. His life provides a good case study of entrepreneurship in the anthracite trade. Like a number of other merchants, he helped establish a market for Pennsylvania anthracite. He was also typical of other merchants who during the decade following the War of 1812 helped change the course of American economic history.

Notes

The following abbreviations appear in the Notes.

ANSP	The Academy of Natural Sciences of Philadelphia
CC	Charles Cist
DRC	Dorrance Reynolds Collection
DS	David Scott
DSC	David Scott Collection
EMHL	Eleutherian Mills Historical Library
FW	Francis Weiss
GHC	George Hollenback Collection
GM	George MacCulloch
GMH	George M. Hollenback
HC	Hollenback Collection
HSP	Historical Society of Pennsylvania
IAC	Isaac Abel Chapman
ICC	Isaac Chapman Collection
JC	Jacob Cist
JCC	Jacob Cist Collection
JN	John Nicholson
JNP	John Nicholson Papers
JR	James Ronaldson
JS	John Smith
JW	Jacob Weiss
JWP	Jacob Weiss Papers
LCMC	Lehigh Coal Mine Company
LCMCP	Lehigh Coal Mine Company Papers
LSC	Lewis S. Coryell
MH	Matthias Hollenback
MHC	Matthias Hollenback Correspondence
PHMC	Pennsylvania Historical and Museum Commission
SS	Smith Section
TWC	Thomas Weiss Correspondence
VMC	Volney Maxwell Collection
WFP	Weiss Family Papers
WHGS	Wyoming Historical and Geological Society

Introduction

1. Thomas Cooper, "Statement on Coal." Found in the papers of the Lehigh Coal Mine Company (hereafter cited as LCMC), Box 1 Shelf 45, Wyoming Historical and Geological Society (hereafter cited as WHGS).

2. Ibid. Thomas Cooper's biography notes his interest in mineralogy but does not mention coal. See Dumas Malone, *The Public Life of Thomas Cooper 1783–1839* (1926; reprint, Columbia, S.C., 1961).

3. Tench Coxe, *A View of the United States, in a Series of Papers, Written At Various Times between the Years 1787 and 1794* (Philadelphia, 1794).

4. Alfred D. Chandler, Jr., "Anthracite Coal and the Beginnings of the Industrial Revolution in the United States," *Business History Review* 46 (1972): 141–81.

5. A few of the popular as well as scholarly works dealing with the Pennsylvania anthracite trade include Anthony J. Bryski, "The Lehigh Canal and Its Effect on the Economic Development of the Region through which It passed, 1818–1873" (Ph.D. diss., New York University, 1957); Eli Bowen, *The Pictorial Sketch-Book of Pennsylvania* (Philadelphia, 1854); Samuel Harris Daddow and Benjamin Bannan, *Coal, Iron, and Oil* (Pottsville, 1866); (The Delaware and Hudson Company), *A Century of Progress: History of the Delaware and Hudson Company 1828–1923* (Albany, 1925); Howard N. Eavenson, *The First Century and a Quarter of American Coal Industry* (Pittsburgh, 1942); (The Hudson Coal Company), *The Story of Anthracite* (New York, 1932); Eleanor Morton, *Josiah White: Prince of Pioneers* (New York, 1946); H. Benjamin Powell, "Coal, Philadelphia, and the Schuylkill" (Ph.D. diss., Lehigh University, 1968); Michael Shegda, "History of the Lehigh Coal and Navigation Company" (Ed.D. diss., Temple University, 1952); Richard Richardson, *Memoir of Josiah White* (Philadelphia, 1873); C. K. Yearly, Jr., *Enterprise and Anthracite: Economics and Democracy in Schuylkill County, 1820–1875* (Baltimore, 1961).

6. Robert G. Albion, *The Rise of New York Port: 1815–1860* (New York, 1939); Robert K. Lamb, "The Entrepreneur and the Community," in *Men in Business: Essays on the Historical Role of the Entrepreneur,* ed. William Miller (1952; reprint, New York, 1962), pp. 91–119.

7. While Robert Albion alludes to this point, it is most forcefully stated in Julius Rubin, "Canal or Railroad? Imitation and Innovation in the Response to the Erie Canal in Philadelphia, Baltimore, and Boston," *Transactions of the American Philosophical Society* n.s. 51, pt. 7 (November 1961).

8. Arthur H. Cole, "An Approach to the Study of Entrepreneurship," and Joseph A. Schumpeter, "Economic Theory and Entrepreneurial History," in *Explorations in Enterprise,* ed. Hugh G. J. Aitken (Cambridge, Mass., 1965), pp. 23–64; Joseph A. Schumpeter, *The Theory of Economic Development* (1934; reprint, New York, 1961), pp. 128–56; David C. McClelland, *The Achieving Society* (1961; reprint, New York, 1967). The concepts of Schumpeter best fit the behavior of Cist.

9. In addition to the works cited in the previous note, see James P. Baughman, "New Directions in American Economic and Business History," in *American History: Retrospect and Prospect,* ed. George A. Billias and Gerald N. Grob (New York, 1971), pp. 301–3; Thomas C. Cochran, *The Inner Revolution: Essays on the Social Sciences in History* (New York, 1964), pp. 55–157; Hugh G. J. Aitken, "The Future of Entrepreneurial Research," *Explorations in Economic History* 2d ser. 1 (Fall 1963): 3–9; Peter F. Bell, "The Direction of Entrepreneurial Explorations: A Review Article," *Explorations in Entrepreneurial History,* 2d ser. 5 (Fall 1967): 3–11; James H. Soltow, "The Entrepreneur in Economic History," *American Economic Review* 58 (May 1968): 84–92.

10. Thomas C. Cochran has developed and used this concept extensively. In addition to his work cited in note 9, see his two articles in Aitken, *Explorations in Enterprise,* pp. 93–112, 123–38, and "The Entrepreneur in Economic Change," *Explorations in Economic*

History, 2d ser. 3 (Fall 1965): 25–38. This concept has been a focal point for his award-winning *Business in American Life: A History* (New York, 1972). See also W. W. Rostow, *The Stages of Economic Growth: A Non-Communist Manifesto* (Cambridge, 1962).

Chapter 1

1. George B. Kulp, *Families of the Wyoming Valley* (Wilkes-Barre, 1890), 3:1336, 1344; *Dictionary of American Biography* (New York, 1929), 2:108–10; Robert K. Lamb, "The Entrepreneur and the Community," in *Men in Business* (New York, 1962), pp. 91–119.

2. Kulp, *Families of the Wyoming Valley,* 3:1332–34.

3. *Columbian,* 2 (1788): 781–86.

4. *Columbian,* 3 (1789): 505; 1 (1787): 868–69.

5. Levin T. Reichel, *A History of Nazareth Hall, from 1755 to 1855* (Philadelphia, 1855); Kulp, *Families of the Wyoming Valley,* 3:1342.

6. Ibid.; Jedidiah Morse, *The American Universal Geography* (Boston, 1789), 1:431–32.

7. Jacob Cist (hereafter cited as JC), "Sketch of Factory, April 13, 1795," Jacob Cist Collection (hereafter cited as JCC), WHGS; Robert D. Arbuckle, "John Nicholson and the Attempt to Promote Pennsylvania Industry in the 1790s," *Pennsylvania History* 42 (April 1975): 99–114; Harold Hutchinson, *Tench Coxe: A Study in American Economic Development* (1938; reprint, New York, 1969), pp. 77–112.

8. JC to Matthias Hollenback (hereafter cited as MH), 29 October, 13 November 1807, JCC, WHGS.

9. *Proceedings of the Lehigh County Historical Society* 21 (1956): 17–18; Kulp, *Families of the Wyoming Valley,* 3:1335–38. See also "Notice of the First Introduction of Anthracite on the Susquehanna," *American Journal of Science* 18 (1830): 307; Jacob Weiss (hereafter cited as JW), "Daybook for Gnadenhutten, 1785–1786," Weiss Family Papers (hereafter cited as WFP), Academy of Natural Science of Philadelphia (hereafter cited as ANSP); Anthony J. Bryski, "The Lehigh Canal and Its Effect on the Economic Development of the Region Through Which It Passed" (Ph.D. diss., New York University, 1957), p. 40.

10. Reading Howell, "Map of Pennsylvania (1792)," *Pennsylvania Archives* 3d ser., Appendix, pp. I–X.

11. Samuel G. Hermelin, *Report About the Mines in the United States of America, 1783* (Philadelphia, 1931), p. 44; Johann D. Schoepf, *Travels in the Confederation (1783–1784)* (Philadelphia, 1911), pp. 175, 205; Gabriel Plattes, *A Discovery of Subterranean Treasure: Containing Useful Explorations Concerning All Manner of Mines and Minerals from the Gold to the Coal* (Philadelphia, 1784), p. 19.

12. Hermelin, *Report About the Mines,* p. 44.

13. Schoepf, *Travels,* p. 169.

14. "Extracts from the Discovery of Subterranean Treasure," Thomas Weiss Correspondence (hereafter cited as TWC), JW Papers (hereafter cited as JWP), Historical Society of Pennsylvania (hereafter cited as HSP).

15. Ibid.

16. Frederick B. Tolles, *Meeting House and Counting House: The Quaker Merchants of Colonial Philadelphia 1682–1763* (1948; reprint, New York, 1963), pp. 95–100.

17. Nonimportation Resolution, 25 October 1765, MS 960, HSP.

18. United States Custom House Records, Inward Entries, 15 March 1784 to 30 April 1785, MS 157, HSP; United States Treasury Department, "Tonnage and Imports for the Year(s) Ending September 30, 1790 (– September 30, 1792)," *American State Papers: Commerce and Navigation* (Washington, D.C., 1832) 1:85, 181, 184, 205.

19. Arbuckle, "John Nicholson," pp. 99–114; Robert D. Arbuckle, *Pennsylvania Speculator and Patriot: The Entrepreneurial John Nicholson, 1757–1800* (University Park, Pa., 1975), pp. 139–64.

20. Diary of John Nicholson (hereafter cited as JN), 1 January to 31 July 1792, John Nicholson Papers (hereafter cited as JNP), Division of Archives, Pennsylvania Historical and Museum Commission (hereafter cited as PHMC).

21. "The First Coal Mining Company of the Lehigh Region," *Pennsylvania Magazine of History and Biography* 39 (1915): 170–75.

22. Bryski, "The Lehigh Canal," p. 70; LCMC, Subscription, Lehigh Coal Mine Company Papers (hereafter cited as LCMCP), 1792–1829, Eleutherian Mills Historical Library (hereafter cited as EMHL).

23. Diary of JN, 1 January to 31 July 1792 (JNP), PHMC; Bryski, "The Lehigh Canal," pp. 69–70.

24. Michael Shegda, "History of the Lehigh Coal and Navigation Company," p. 16.

25. Diaries of JN, 1792–1794, JNP, PHMC; Arbuckle, *Entrepreneurial John Nicholson.* See also Norman B. Wilkinson, "Land Policy and Speculation in Pennsylvania, 1779–1800" (Ph.D. diss., University of Pennsylvania, 1958).

26. Francis Weiss (hereafter cited as FW) to Josiah White, 2 November 1826, LCMCP, EMHL; Shegda, "History of the Lehigh Coal and Navigation Company," p. 19.

27. *Encyclopedia; or a Dictionary of Arts, Sciences and Miscellaneous Literature* (Philadelphia, 1798), 5:87–105; "Coal Mines Method of Working from the Encyclopedia—with Additional Notes by T. Bedwell," LCMC, WHGS.

28. Diary of JN, 1 January to 31 July 1792, JNP, PHMC.

29. Shegda, "History of the Lehigh Coal and Navigation Company," p. 18; JC, "Account of the Mines of Anthracite in the Region about Wilkes-Barre," *American Journal of Science* 4 (1822): 11–12; Charles Cist (hereafter cited as CC) to FW, Philadelphia, 14 July 1801, TWC, JWP, HSP.

30. Diary of JN, 1 January to 31 July 1792, JNP, PHMC.

31. "Coal Mines Method of Working from the Encyclopedia," LCMC, WHGS; G. Haga to FW, Philadelphia, 13 June 1798, JWP, HSP.

32. Thomas C. James, "A Brief Account of the Discovery of Anthracite Coal on the Lehigh," *Memoirs of the Historical Society of Pennsylvania* 1 (1826): 325–26.

33. *Susquehanna Democrat,* Wilkes-Barre, 25 December 1818.

34. Shegda, "History of the Lehigh Coal and Navigation Company," p. 17; William Henry to JN, Nazareth, 28 September 1792, JNP, PHMC.

35. Shedga, "History of the Lehigh Coal and Navigation Company," p. 18; FW to CC, 1 August 1800, CC to FW, Philadelphia, 14 July 1801, TWC, JWP, HSP; G. Haga to JN, Philadelphia, 5 January 1796, JNP, PHMC.

36. FW to CC, 1 August 1800, FW Correspondence, JWP, HSP.

37. CC to FW, Philadelphia, 14 July 1801, TWC, JWP, HSP; Tench Coxe, *A View of the United States,* p. 71.

38. Leonard W. Labaree and Whitefield J. Bell, eds., *Papers of Benjamin Franklin* (New Haven, 1960), 2:419–45; Josephine H. Peirce, *Fire on the Hearth: The Evolution and Romance of Heating Stoves* (Springfield, Mass., 1951), pp. 39–74; Jared Sparks, ed., *The Works of Benjamin Franklin* (Boston, 1838), 4:543–60.

39. "Oliver Evans on His Stove for Burning Lehigh Coal," *Archives of Useful Knowledge* 2 (1811): 291–93.

40. JN to Michael Hillegas, Philadelphia, 4 April 1796, JN Letterbook IV, HSP.

41. Kulp, *Families of the Wyoming Valley,* 3:1335.

42. JC to MH, Washington, D.C., 13 November 1807, JCC, WHGS; JC to MH, Washington, D.C., March 1808, JCC, WHGS; List of the Stockholders 1814, JCC, ANSP.

43. Agreement between MH and JC, 13 February 1808, JCC, WHGS.

44. Jacob Eyerly, Jr., to MH, Nazareth, 6 April, 2 September 1792, MHC, HC, WHGS; Pennsylvania Land Office, "A Connected Warrantee Township Map of Hazle, Luzerne County." See also William J. Buck, "History of the Early Discovery of Coal," *Potter's American Monthly* 4 (1875): 180–82; Pennsylvania Land Office, Connected Warrantee Township Maps for Luzerne County.

45. Howard Eavenson, *The First Century and a Quarter of American Coal Industry* (Pittsburgh, 1942), pp. 49, 51; JC to MH, Washington, D.C., 13 November 1807, JCC, WHGS.

46. JC to MH, Washington, D.C., 13 November 1807, 19 March 1808, JCC, WHGS.

47. JC to MH, Washington, D.C., 13 November 1807, JCC, WHGS.

48. Kulp, *Families of the Wyoming Valley*, 3:1344.

49. "On Coal as a Fuel in America," *Literary Magazine* 2 (1804): 422–24.

50. Kulp, *Families of the Wyoming Valley*, 1:347; Erskine Hazard, "History of the Introduction of Anthracite Coal into Philadelphia" and "A Letter from Jesse Fell of Wilkes-Barre on the Discovery and First Use of Anthracite into the Wyoming Valley," *Memoirs of the Historical Society of Pennsylvania* 2 (1827): 155–65.

51. Kulp, *Families of the Wyoming Valley*, 2:784.

52. Charles Francis Richardson, "Reminiscences of Hon. Charles Miner, 1780–1865," *Proceedings and Collections of the Wyoming Historical and Geological Society* 14 (1915): 86–87, 110–15.

53. JC, Catalogue of Books, JCC, WHGS.

54. Venia Phillips is currently working on a biography of Cist with special reference to his work in entomology.

55. Benjamin Franklin, *Autobiography* (New Haven, 1964), p. 172.

56. Articles of Association of the Lehigh Coal and Lumber Company (1815), JCC, WHGS.

57. JC died on 30 December 1825 at age 43, and MH died on 18 February 1829 at age 77.

Chapter 2

1. James Ronaldson (hereafter cited as JR) to JC, Philadelphia, 13 March 1813, JCC, ANSP; Arthur H. Cole, *Wholesale Commodity Prices in the United States 1700–1861: Statistical Supplement Actual Wholesale Prices of Various Commodities* (Cambridge, 1938).

2. Edwin Wolf II, *Philadelphia: Portrait of an American City* (Harrisburg, 1975), p. 126.

3. *United States Gazette*, 20 May 1815; Greville Bathe and Dorothy Bathe, *Oliver Evans: A Chronicle of Early American Engineering* (Philadelphia, 1935), p. 196.

4. Statistics for this early period are not completely accurate. See Tench Coxe, *A Statement of the Arts and Manufactures of the United States of America for the Year 1810* (Philadelphia, 1814), pp. 44–74.

5. Oliver Evans, "Oliver Evans and His Stove for Burning Lehigh Coal," *Archives of Useful Knowledge* 2 (1812): 291–93; Bathe and Bathe, *Oliver Evans*, pp. 195–96.

6. Philadelphia Water Committee, *Report(s) of the Watering Committee to the Select and Common Councils*, 7 October 1805; 13 November 1806; 13 November 1807; 24 November 1808; Watering Committee Papers, Archives of the City and County of Philadelphia, City Hall, Philadelphia; see also Benjamin Henry Latrobe, "First Report of Benjamin Henry Latrobe to the American Philosophical Society," *Transactions of the American Philosophical Society* 6 (1809): 89–98. I would like to thank Darwin Stapleton for pointing out this information to me.

7. United States Treasury Department, "Imports for the Year(s) Ending September 30, 1790 (– September 20, 1822)," *American State Papers: Commerce and Navigation,* 2 vols. (Washington, D.C., 1832).

8. *United States Gazette,* 20 May 1815.

9. JR to JC, Philadelphia, 13 March and 30 June 1813, JCC, ANSP.

10. Ibid., 7 July and 11 August 1813, John Stover Receipt, Wilkes-Barre, 9 July 1813, George Sively Receipt, Wilkes-Barre, 5 August 1813, JCC, ANSP.

11. William L. Brockelman to JC, Baltimore, 12 April 1813, JCC, ANSP.

12. Joshua Gilpin, "Journal of a Tour from Philadelphia Thro the Western Counties of Pennsylvania in the Months of September and October, 1809," *Pennsylvania Magazine of History and Biography* 50 (1926): 76–77.

13. William L. Brockelman to JC, Baltimore, 1, 4, 31 May, 26 July 1813, JCC, ANSP.

14. W. L. Brockelman to JC, Philadelphia, 21 April 1812, JCC, ANSP.

15. Joshua Malin to Gerard Ralston, Lebanon, 20 April 1827, Archives, HSP.

16. Isaac A. Chapman (hereafter cited as IAC) to JC, Springville, Susquehanna County, May 1812, JCC, ANSP; agreement between JC and IAC, 12 July 1812, JCC, WHGS.

17. JC, Notes on LCMC, 1813, JCC, ANSP; JC to IAC, Wilkes-Barre, 14 March 1813, JCC, WHGS.

18. "Statement of the probable expense of taking coal from the Lehigh Coal Mine," JCC, ANSP.

19. Indenture between LCMC and JC, Charles Miner and John Robinson, 15 December 1813, JCC, ANSP.

20. Agreement between LCMC and JC, 15 December 1813, JCC, ANSP.

21. John Robinson to JC, Lausanne, 21 February and 19 March 1814, JCC, ANSP.

22. Ibid.; No. 1 Castle Building, 25 March 1814, JCC, WHGS.

23. Memorandum of Agreement between JC and Nathan Miner, 4 March 1814, JCC, ANSP; account of Nathan Miner, LCMC, Account Book 1814–1815, JCC, ANSP.

24. John Robinson to JC, Lausanne, 19 March 1814, JCC, ANSP.

25. JC to G. Haga (undated), Cist Section, Volney Maxwell Collection (hereafter cited as VMC), WHGS.

26. No. 1 Castle Building, 25 March 1814, JCC, WHGS.

27. Pennsylvania Senate, "Report of the Committee of the Senate, upon the Subject of the Coal Trade, March 4, 1834 (Samuel Packer Chairman)," *Journal* (Harrisburg, 1833–34) 2:540 (hereafter cited as Packer Report).

28. Account of Nathan Miner, LCMC, Account Book, 1814–1815, JCC, ANSP.

29. Benjamin Silliman, "Notes on a Journey from New Haven, Conn., to Mauch Chunk and Other Anthracite Regions of Pennsylvania," *American Journal of Science, and Arts* 19 (1831): 20.

30. JC, Advertisement, Germantown, 15 July 1814, JCC, WHGS.

31. Account of Coal, Account of Isaac Backman, LCMC, Account Book, 1814–1815, JCC, ANSP.

32. Contract between JC and Manna and Jo Thompson, 12 December 1814, microfilm, ANSP.

33. (IAC), Statement, 1815, JCC, ANSP.

34. See for example Martin Herter Receipt, 15 February 1815, JCC, ANSP; IAC to JC, Lausanne, 6 March 1815, Isaac Chapman Collection (hereafter cited as ICC), WHGS.

35. Israel D. Rupp, *History of Northampton, Lehigh, Monroe, Carbon, and Schuylkill Counties* (Harrisburg, 1845); IAC, Town of Lehighton, ICC, WHGS.

36. Joseph Horsfield to JC, Bethlehem, 12 April 1814, JCC, ANSP.

37. Nathan Miner, Memorandum, 8 August 1814, JCC, ANSP.

38. Account of Coal, Account of Michael Frey at Easton, LCMC, Account Book, 1814–15, JCC, ANSP.

39. J. J. and William Henry to JC, Boulton, 12 August 1814, Account of J. J. and William Henry, LCMC, Account Book, 1814–15, JCC, ANSP.

40. JC, Handbill, 15 July 1814, JCC, WHGS.

41. Account of Joseph Smith, LCMC, Account Book, 1814–15, JCC, ANSP; Cist, "Account of the Mines of Anthracite," pp. 13–14.

42. Packer Report, 540.

43. Memorandum of Agreement between JC and IAC, 12 July 1814, JCC, ANSP.

44. Nathan Miner, Memorandum, 8 August 1814, JCC, ANSP; account of Abiel Abbott, LCMC, Account Book, 1814–15, JCC, ANSP.

45. Packer Report, 540.

46. Ibid., pp. 540–41.

47. IAC, Statement, 1815, JCC, ANSP; Cist, "Account of the Mines of Anthracite," pp. 1–16.

48. Miner's account of the Lehigh coal trade was written twenty years later in defense of monopolistic charges against the Lehigh Coal and Navigation Company. His bias reflected conditions in 1833. Most of his facts were accurate. Packer Report, pp. 539–42.

49. IAC, Statement, 1815, JCC, ANSP.

50. Ibid.

51. Ibid.; Memorandum of an Agreement between JC and Stephen Tuttle and Lyman Covell, 6 September 1814, JCC, ANSP.

52. LCMC, Resolution, 31 October 1814, JCC, ANSP.

53. Joshua Whitney to JC, Binghamton, 7, 9 November 1814, JCC, ANSP.

54. IAC to JC, Lausanne, 27 November 1814, JCC, ANSP; IAC to JC, Binghamton, 5 December 1814, JCC, ANSP.

55. Henry D. Mandeville to JC, Allentown, 4 January 1815, JCC, WHGS.

56. Packer Report, 541; IAC, Statement, 1815, JCC, ANSP; Steelwagon and Knight to JC, Philadelphia, 7 December 1814, microfilm, ANSP.

57. Steelwagon and Knight to JC, Philadelphia, 15 February 1815, JCC, ANSP.

58. IAC, Statement, 1815; Tuttle and Covell in Account current with JC, JCC, ANSP.

59. Charles Miner to JC, New York, 14 November 1814, Charles Miner Collection, WHGS.

60. JC (Notes on potential New York coal merchants), Cist Section, VMC, WHGS; JC to Joshua Barker & Co., Wilkes-Barre, 3 March 1815, JCC, WHGS.

Chapter 3

1. (IAC), Statement, 1815, JCC, ANSP; Pennsylvania Senate, "Report of the Committee of the Senate, upon the Subject of the Coal Trade, March 4, 1834 (Samuel Packer, chairman)," *Journal* 2 (Harrisburg, 1833–34): 541.

2. JC, "On the Importance and Necessity of Introducing Coal, Particularly the Species Known by the Name of Lehigh Coal into Immediate and General Use," JCC, ANSP; JC, "Account of the Mines of Anthracite in the Region about Wilkes-Barre, Pennsylvania," *American Journal of Science* 4 (1822): 1–16; JC to IAC, Wilkes-Barre, 7 April 1815, JCC, WHGS.

3. Ibid.; *The Gleaner*, 24 March 1815; *Facts Illustrative of the Character of the Anthracite, or Lehigh Coal, Found in the Great Mines at Mauch Chunk, in Possession of the Lehigh Coal and Navigation Company, with Certificates from Various Manufacturers Proving Its Decided Superior-*

ity Over Every Other Kind of Fuel (Boston: T.R. Marvin, 1825), reprinted in John N. Hoffman, "Anthracite in the Lehigh Region of Pennsylvania, 1820–1845," in *Contributions from the Museum of History and Technology* (Washington, D.C., 1968), pp. 132–41.

4. "A Series of Talks on the Several Branches of American Manufactures Exhibiting Them in Every County of the Union (1810)," in *A Statement of the Arts and Manufactures of the United States of America for the year 1810* (Philadelphia, 1814), p. 52 (hereafter cited as County Series, *Statement of the Arts*); JC, Memoranda at Philadelphia and on the road, 1814, JCC, ANSP.

5. Joseph Smith, Testimonial, 23 February 1814, JCC, ANSP.

6. *United States Gazette,* 22 May 1815.

7. Joseph Smith to JC, Tinicum, 4 April 1815, JCC, WHGS.

8. County Series, *Statement of the Arts,* p. 50.

9. George Smith et al. to JC, Eagle Nail Factory, 7 December 1814, JCC, ANSP.

10. *United States Gazette,* 22 May 1815.

11. Erskine Hazard, "History of the Introduction of Anthracite Coal into Philadelphia," *Memoirs of the Historical Society of Pennsylvania* 2 (1827): 159–60.

12. "Schuylkill Coal," *American Daily Advertiser,* 1814, LCMC, WHGS; JC, "Account of the Mines," p. 10; *United States Gazette,* 22 May 1815.

13. *United States Gazette,* 22 May 1815.

14. White and Hazard to JC, Philadelphia, 18 March 1815, JCC, WHGS.

15. JC, Notes, JCC, ANSP.

16. *United States Gazette,* 22 May 1815.

17. Ibid.; JC, "Account of the Mines," p. 8.

18. William Hood, Statement on Coal, JCC, ANSP; Bathe and Bathe, *Oliver Evans,* pp. 122–23.

19. *United States Gazette,* 22 May 1815.

20. James Woodhouse, "Lehigh Coal," *Hazard's Register of Pennsylvania* 16 (1835): 225–26.

21. Designs of glass furnaces, JCC, ANSP.

22. "Stone Coal," *The Emporium of Arts and Sciences* 2 (1813): 477; Woodhouse, "Lehigh Coal," pp. 225–26.

23. Oliver Evans to JC, Washington, D.C., 3 January 1815, JCC, ANSP; JC, "Account of the Mines," p. 15.

24. Oliver Evans to JC, Washington, D.C., 3 January 1815, JCC, ANSP.

25. IAC to James Mease, Wilkes-Barre, 20 March and 1 May 1810, Miscellaneous Communication to the American Philosophical Society, Mechanics Machinery Engineering, 1:36, 366, 506.73 Am4mc; IAC, "Illustration of a Steam Boat 1810," ibid.; Robert Patterson, Statement Concerning Isaac Chapman's Steamboat, 27 July 1810, ibid., American Philosophical Society.

26. James Eckroyd, "On the Use of Lime in England," *Memoirs of the Philadelphia Society for Promoting Agriculture* 3 (1814): 37–41.

27. County Series, *Statement of Arts,* pp. 73–74.

28. Ibid., pp. 58–59.

29. JC, "Account of the Mines," pp. 8–9, 12–13.

30. Ibid.

31. James Mease, *The Picture of Philadelphia* (Philadelphia, 1811), pp. 125–28.

32. JC, "Account of the Mines," pp. 8–9.

33. JC, Notes, Cist Section, VMC, WHGS; JC, Notes, JCC, ANSP.

34. JC, "Account of the Mines," pp. 10–11.

35. James, "A Brief Account of the Discovery of Anthracite Coal on the Lehigh," *Memoirs of the Historical Society of Pennsylvania* 1 (1826): 328; *United States Gazette,* 22 May 1815.

36. White and Hazard to JC, Philadelphia, 18 March 1815, JCC, WHGS.

37. JC, "For the American Daily Advertiser," JCC, ANSP.

38. Ibid.

39. (JC), Copy of letter to someone about introducing coal into Russia, Cist Section, VMC, WHGS.

Chapter 4

1. JC to IAC, Wilkes-Barre, 7 April 1815, JCC, WHGS.

2. JC, List of the 1814 Stockholders, JCC, ANSP.

3. William Henry and JC, Memorandum of an Agreement, Nazareth, 27 November 1814, ibid., p. 561; William Henry to JC, Bethlehem, 4 January 1815, JCC, WHGS; John F. Stadiger Receipt, Bethlehem, 25 December 1814, Lord Butler Collection, WHGS.

4. JC, Sketch of Lehigh Coal Mine, Cist Section, VMC, WHGS.

5. JC and IAC, Memorandum of a Lease, Lausanne, 23 April 1815, JCC, WHGS.

6. IAC to JC, Springville, 26 December 1814, Charles Fisher Wells Collection, microfilm, ANSP.

7. JC, Wilkes-Barre, 14 March 1815, JCC, WHGS.

8. IAC to JC, Springville, 8 January 1815, ICC, WHGS.

9. William Thomas to JC, Philadelphia, 11 March 1815, JCC, WHGS.

10. White and Hazard to JC, Philadelphia, 18 March 1815, JCC, WHGS.

11. *The Gleaner,* 17 March 1815.

12. IAC to JC, Lausanne, 6 April 1815, ICC, WHGS.

13. IAC to JC, Lausanne, 31 March and 6 April 1815, ICC, WHGS; JC to IAC, Wilkes-Barre, 2, 6 May 1815, IAC to JC, Lausanne, 9 May 1815, JCC, WHGS.

14. IAC to JC, Lausanne, 31 March 1815, ICC, WHGS.

15. JC to IAC, Wilkes-Barre, 7 April 1815, JCC, WHGS.

16. Ibid.

17. Joseph Smith to JC, Tinicum, 4 April 1815, JCC, WHGS.

18. Ibid.

19. Joseph Smith to Cist and Miner, Doylestown, 3 May 1815, JCC, WHGS.

20. JC to IAC, Wilkes-Barre, 6 May 1815, JCC, WHGS.

21. LCMC, Account Book 1814–1815, JCC, ANSP; JC to IAC, Wilkes-Barre, 2 May 1815, JCC, WHGS.

22. JC to IAC, Wilkes-Barre, 17 June 1815, JCC, WHGS; JC to MH, Wilkes-Barre, 5 July 1815, HC, WHGS.

23. Articles of Association of the Lehigh Coal and Lumber Company, 1815, JCC, WHGS.

24. Ibid.

25. Asa Morris, Jr., to ———, New Haven, 19 September 1815, Stuben Butler correspondence, misc. file, WHGS.

26. IAC to JC, Lausanne, 25 December 1815, ICC, WHGS.

27. Ibid.

28. JR to JC, Philadelphia, 22 December 1815, JCC, WHGS.

29. IAC to JC, Lausanne, 26 May 1816, Charles Fisher Wells Collection, microfilm, ANSP; Robinson-Chapman and Co., Day Book 1815–1820, JCC, WHGS.

30. *The Gleaner,* 15 August 1817; Charles F. Richardson, "Reminiscences of Hon. Charles Miner, 1780–1865," *Proceedings and Collections of the Wyoming Historical and Geological Society* 14 (1915): 127–35.

31. *Susquehanna Democrat,* 25 December 1818.

32. JW, Jr., to JC, Drumhellers, 7 January 1818, LCMCP, WHGS.

33. For White's biography see Josiah White, *Journal Given by Himself* (Philadelphia, 1909); Richard Richardson, *Memoir of Josiah White* (Philadelphia, 1873); Eleanor Morton, *Josiah White* (New York, 1946).

34. Richardson, *Memoir of Josiah White,* pp. 20–21.

35. Josiah White et al., Original Draft for Bill Improving the River Schuylkill, 1812, ALS, HSP; H. Benjamin Powell, "Coal, Philadelphia and the Schuylkill," pp. 94–112.

36. *Weekly Advertiser,* Reading, 14 October 1815; Schuylkill Navigation Company, "Minutes of the Managers, Book A," MG 110, PHMC; Pennsylvania Laws, *Acts of the Legislature of Pennsylvania Relative to the Schuylkill Navigation Company* (Philadelphia, 1869), p. 14.

37. Schuylkill Navigation Company, "Minutes of the Managers, Book A," 21 August 1818, 21 October 1818, 5, 20, 25 January 1819, MG 110, PHMC; Philadelphia Watering Committee, *An Additional Report on Water Power* (Philadelphia, 1819).

38. The act of incorporation in 1815 gave the Schuylkill Navigation Company the right to raise $500,000 in capital stock. Subsequent legislation permitted the company to increase this amount. By January 1826 the managers had paid $1,808,054 for improvements, land, and damages. Schuylkill Navigation Company, *Report of the President and Managers . . . to the Stockholders, January 2, 1826* (Philadelphia, 1826), p. 10.

39. White, *Journal,* pp. 17–20.

40. Michael Shegda, "History of the Lehigh Coal and Navigation Company," pp. 44–47.

41. IAC Journal, in Alfred Mathews and Austin N. Hungerford, *History of the Counties of Lehigh and Carbon, in the Commonwealth of Pennsylvania* (Philadelphia, 1884), p. 661.

42. John N. Hoffman, "Anthracite in the Lehigh Region of Pennsylvania, 1820–1845," *United States National Museum Bulletin* 252 (Washington, D. C., 1968): 130–31.

43. White, *Journal,* pp. 22–26.

44. Lehigh Navigation Company, Terms of Subscription, 10 August 1818; Lehigh Coal Company, Articles of Agreement, 19 October 1818; Lehigh Navigation and Coal Company, Articles of Agreement, 21 April 1820; LCMC, Articles of Agreement with Josiah White and Erskine Hazard, 1 May 1821, Lehigh Coal and Navigation Company Main Office, Bethlehem.

45. White, *Journal,* pp. 19, 33–39; Shegda, "History of the Lehigh Coal and Navigation Company," pp. 61–65.

46. FW to JC, Toamensin, 8 October 1820, JCC, ANSP.

47. Frederick Steelwagon to JC, Philadelphia, 31 December 1821, JCC, WHGS.

Chapter 5

1. See Louis E. Atherton, "The Pioneer Merchant in Mid-America," *University of Missouri Studies* 14 (April 1939); Louis E. Atherton, *The Southern Country Store, 1800–1860* (Baton Rouge, La., 1949).

2. GMH to M. H. Laning, 18 October 1834, GMH Memorandum Book 1823–1840, HC, WHGS.

3. Luke Sarsfield, "Matthias Hollenback: Frontier Merchant" (Ed.D. diss., New York University, 1972); see MH, Pocket Memorandum Book, 1813–1819, 1819–1825, HC, WHGS.

4. United States Bureau of Census, *Fifth Census of the United States* (Washington, D.C., 1840), pp. 64–67.

5. "Map of the Wyoming and Lackawanna Valleys," *American Journal of Science* 18 (1830): 328.

6. Accounts of Thomas Borbridge & Company, Henderson Gaylord, and Benjamin Reynolds, JS, Account Book 1814–1839, Coal Book 1818–1824, DRC, WHGS.

7. Compiled from JS, Account Book 1814–1839, Coal Book 1818–1824, DRC, WHGS.

8. Ibid.; MH and JC, Memorandum Book 1809–1824, HC, WHGS; Samuel Hodgson to MH, Harrisburg, 19 January 1815, HC, WHGS; Paul W. Gates, *The Farmer's Age: Agriculture, 1815–1860* (New York, 1960), p. 55.

9. Henderson Gaylord to GMH, Plymouth, 10 February 1825, HC, WHGS.

10. John Gibson to GMH, 22 October 1817, HC, WHGS.

11. S. Henry to GMH, Cactariw Furnace, 31 October 1826, HC, WHGS.

12. Reuben Trexler to MH, Mary Ann Furnace, 2, 15 December 1818, 28 September 1821, HC, WHGS.

13. MH and JC, Memorandum Book 1809–1824, HC, WHGS.

14. William H. Alexander to GMH, Berwick, 16 October, 24 November 1828, HC, WHGS; Patrick Hepburn to GMH, Montrose, 18 September 1830, HC, WHGS.

15. JS, Account Book 1814–1839, Coal Book 1818–1824, DRC, WHGS.

16. MH and JC, Memorandum Book 1809–1824 (Entries dated 14 December 1815, 20, 25 October 1819, 24 January, 2 February 1820, 18 November 1822), HC, WHGS; GMH, Memorandum Book, 1823–1840 (Entries dated 28 October 1826, 14 February 1828), HC, WHGS.

17. MH and JC, Memorandum Book 1809–1824, GMH, Memorandum Book 1823–1840, HC, WHGS.

18. JS, Account Book 1814–1839, Coal Book 1818–1824, DRC, WHGS; see accounts of Ira Brownson, Adin Linn, in Account Book 1814–1839, and Harry Gabriel and Joseph Walton in Coal Book 1818–1824.

19. Accounts of Samuel Actly and Aron Clawson, Account Book 1814–1839, Accounts of James Handle, James Hodge, Ebenezer Marble, and John Vanloon, Coal Book 1818–1824.

20. Account of Anson Martin, Coal Book 1818–1824.

21. Accounts of Thomas Pringle, John Turner, and Noah Wadhanis, Coal Book 1818–1824; Accounts of Henderson Gaylord and Joseph Wright, Coal Book 1818–1824.

22. JS, Account Book 1814–1839, Coal Book 1818–1824, SS, DRC.

23. Ibid.

24. Ibid.

25. Henderson Gaylord to David Scott (hereafter cited as DS), Plymouth, 29 February 1828, David Scott Collection (hereafter cited as DSC), WHGS.

26. Freeman Hunt, ed., *Merchants Magazine* 40 (1859): 246; Schuylkill Navigation Company, *Report of the President and Managers* (Philadelphia, 1828), p. 9.

27. GMH, Memorandum Book 1823–1840, HC, WHGS.

28. *The Gleaner,* 31 May 1816.

29. Ibid., 11 April 1817; MH and JC, Memorandum Book 1809–1824, HC, WHGS.

30. MH and JC, Memorandum Book 1809–1824, HC, WHGS.

31. *United States Gazette,* Philadelphia, 4 May 1804.

32. *The Gleaner,* 20 May 1818.

33. Zachariah [*sic*] Cist, "Account of the Mines of Anthracite, in the Region about Wilkes-Barre, Pennsylvania," *American Journal of Science* 4 (1822): 2; JC, "Fossil Reliquiae of the Anthracite Formation of Pennsylvania," LCMCP, WHGS.

34. J. A. Clark, *The Wyoming Valley* (Scranton, 1875), p. 99; *Susquehanna Democrat,* 22 October 1819; JC, "Account of the Mines of Anthracite," pp. 6–7.

35. Ibid.

36. JC, "Sketch Book Coal Mines Luzerne with Notes Made on the Spot," JCC, ANSP.

37. *Susquehanna Democrat,* 28 February 1823.
38. JC, "Sketch Book Coal Mines."

Chapter 6

1. Thomas P. Roberts, *Memoirs of John Bannister Gibson, Late Chief Justice of Pennsylvania* (Pittsburgh, 1890), p. 45; George B. Kulp, *Families of the Wyoming Valley,* 3: 1091.
2. JC, "Sketch Book Coal Mines."
3. Ibid.
4. In 1821 Cist said that the largest vein was 30 feet in diameter, which is the average size of the seam to become famous as the Mammoth Vein. "Fossil Reliquiae of the Anthracite Formation of Pennsylvania."
5. Ibid.; JC, "Mineralogical Notice of Lehigh Coal," JCC, ANSP.
6. JC, "Fossil Reliquiae of the Anthracite Formation of Pennsylvania."
7. Samuel B. Fisher, *Map of the First and Second Anthracite Fields* (Pottsville, 1836); Henry D. Rogers, *Geological and Topographical Map of the Anthracite Fields of Pennsylvania* (Philadelphia, 1858).
8. JC, The Wyoming Valley in Luzerne County Pennsylvania, JCC, ANSP.
9. JC, "Form of the Luzerne Coal Formation," JCC, WHGS; JC, "Account of the Mines of Anthracite, in the region about Wilkes-Barre, Pennsylvania," p. 4.
10. JC, A profile or cross section of Wyoming Valley about one mile above the Nanticoke Dam, JCC, ANSP.
11. Northern Coal Field Mine Sheet, no. 3, Nanticoke, *Second Geological Survey of Pennsylvania.*
12. See Hugh G. J. Aitken, "A New Way to Pay Old Debts, A Canadian Experience," in *Men in Business: Essays on the Historical Role of the Entrepreneur,* ed. William Miller (1952; reprint, New York, 1962), pp. 71–90; Richard C. Wade, *The Urban Frontier: Pioneer Life in Early Pittsburgh, Cincinnati, Lexington, Louisville, and St. Louis* (1959; reprint, Chicago, 1964).
13. *American Journal of Science* 1 (1818): 1–8.
14. *Dictionary of Scientific Biography,* s.v. "Brongniart, Alexandre," ed. Charles C. Gellespie (New York, 1970), 2:493–97; *American Journal of Science* 1 (1818): 71–74.
15. JC, "Account of the Mines of Anthracite, in the Region about Wilkes-Barre, Pennsylvania," pp. 1–16.
16. Alexandre Brongniart, *Histoires des Végétaux Fossiles* (Paris, 1828) 1, plates XX, LXX, ciii, CXL. I wish to thank William C. Darrah for bringing this material to my attention.
17. *American Journal of Science* 4 (1822): 38–39; 9 (1825): 165–66.
18. Dumas Malone, "Thomas Cooper," *Dictionary of American Biography* (New York, 1929), 2: 414–16; *Susquehanna Democrat,* 1 January 1819.
19. John B. Gibson to JC, Philadelphia, 7 April 1821, JCC, ANSP.
20. James Renwick to JC, 12 August and 13 December 1824, JCC, ANSP.
21. Nicholas Biddle to JC, Philadelphia, 6 August 1823, microfilm, Charles Fisher Wells Collection, ANSP.
22. JC to John P. De Gruchey, Wilkes-Barre, 3 January 1820, Russell Collection, HSP.
23. Ibid.
24. JC, Drought connecting Lands of William Russell, Roberts Vaux, and the Estate of George Roberts . . . Situate in the County of Luzerne, Pennsylvania, Map Case, HSP; Precis of Lackawanna Land, Russell Collection, HSP.

25. JC, Map of Part of the Luzerne Coal Formation Showing the Waters, Owners of Land, Turnpike, State, and a Few of the Common Roads, WHGS.

26. *The Gleaner,* 4 October 1816.

27. IAC, Map of Luzerne County, Pennsylvania (1816), Pennsylvania Land Office, Harrisburg.

28. IAC, Map of Northampton County, Pennsylvania (1817), ibid.

29. Account with North American Land Company 1818–1824; Resurvey of the North American Land Company's land on the Lehigh, 1822; Draft of the North American Company's land, ICC, WHGS; see also Thomas F. Wharton et al., Articles of Agreement, 6 March 1837, Gratz Collection, HSP.

30. Jeddiah Irish, Survey of Nine Tracts of Land in Luzerne and Northampton Counties west of the Lehigh in June and July, 1820, ibid.

Chapter 7

1. H. Benjamin Powell, "Coal, Philadelphia, and the Schuylkill," pp. 94–132.

2. For a general discussion of this movement, see James W. Livingood, *The Philadelphia-Baltimore Trade Rivalry 1780–1860* (Harrisburg, 1947).

3. *Susquehanna Democrat,* 9 January 1818.

4. Ibid., 16 January 1818.

5. *The Gleaner,* 20 May 1818.

6. *Susquehanna Democrat,* 3 July 1818, 1 January and 16 April 1819, 19 October 1821.

7. Ibid., 16 February 1821, 12 April 1822.

8. Ibid., 21 December 1821.

9. Ibid.

10. Ibid., 2 August 1822.

11. Ibid.

12. Ibid., 25 October 1822.

13. Ibid., 15 November 1822.

14. Newspaper clippings, JCC, ANSP.

15. *Susquehanna Democrat,* 3, 31 January 1823; *Wyoming Herald,* Wilkes-Barre, 4, 11 February, 4, 25 March 1825; "Internal Improvements," Broadside, 12 February 1825, Meredith Collection, WHGS.

16. *Wyoming Herald,* 11 February 1825.

17. Richard Shelling, "Philadelphia and the Agitation in 1825 for the Pennsylvania Canal," *Pennsylvania Magazine of History and Biography* 62 (April 1938): 175–77.

18. Petitions to the Pennsylvania State House and Senate, House File, 49th Session – 1, 1824-1825, RG 7, PHMC.

19. Pennsylvania House of Representatives, *Journal of the House of Representatives, 1824–1825* (Harrisburg, 1825) 1:3–4; Pennsylvania Senate, *Journal of the Senate, 1824–1825* (Harrisburg, 1824) 3–4; for an excellent account of Pennsylvania's sectional diversity and political factionalism during this decade, see Philip S. Klein, *Pennsylvania Politics 1817–1832: A Game without Rules* (Philadelphia, 1940).

20. DS to GMH, Wilkes-Barre, 17, 26 February, 16, 25 March 1825, GMH Section, HC, WHGS.

21. Richard Peters, Jr., to JC, Philadelphia, 14 February 1825, JCC, ANSP.

22. H. Benjamin Powell, "Coal, Philadelphia and the Schuylkill."

23. Petitions to the Pennsylvania State House and Senate, House File, 49th Session – 1, 1824-1825, RG 7, PHMC.

24. *Wyoming Herald,* 1 April 1825.

25. Ibid., 15 April 1825; *United States Gazette,* 18 March 1825.

26. Jason Torrey to Lewis S. Coryell (hereafter cited as LSC), Bethany, 7 January, 13 April 1815, LSC Letterbook, I, 24, 26, HSP.

27. Samuel Preston to LSC, Stockport, 26 March, 15 October 1818, 14 September 1823, ibid., I, 42, 51, 100; Jonathan Fell, president of the LCNC, to LSC, Philadelphia, 22 September 1822, ibid., I, 90.

28. DS to LSC, Wilkes-Barre, 12 September 1820, 29 October, 22 December 1823, ibid., I, 67, 101, 103.

29. Samuel Preston to LSC, 14 September 1823, ibid., I, 100.

30. Josiah White to LSC, Mauch Chunk, 28 January 1824, ibid., I, 105; Jacob Wagner to LSC, Easton, 5 February 1824, ibid., I, 107.

31. Petitions to the Pennsylvania State House and Senate, House File, 49th Session – 1, 1824–1825, RG 7, PHMC; *Wyoming Herald,* 28 January 1825.

32. Richard Peters, Jr., to JC, Philadelphia, 14 February 1825, JCC, ANSP.

33. *United States Gazette,* Philadelphia, 22 March 1825.

34. *United States Gazette,* 18, 22 March 1825; *Wyoming Herald,* 25 March, 15 April 1825; *Lancaster Intelligencer,* 29 March 1825.

35. Richard Peters, Jr., to Mathew Carey, 3 May 1825, Edward Carey Gardener Collection, HSP; Shelling, "Philadelphia and the Agitation in 1825 for the Pennsylvania Canal," p. 181.

36. DS to GMH, Wilkes-Barre, 17, 26 February, 16, 25 March 1825, GMH Section, HC, WHGS.

37. *United States Gazette,* 23 September 1825.

38. Avard Bishop, "The State Works of Pennsylvania," *Transactions of the Connecticut Academy of Arts and Sciences* 13 (1907): 265; Powell, "Coal, Philadelphia and the Schuylkill," pp. 94–132. Figures on the number of delegates vary between 113 and 123. Bishop lists 123 as taken from the *United States Gazette,* 12 August 1825.

39. *Pittsburgh Gazette,* 12 August 1825; *Wyoming Herald,* 12 August 1825.

40. Bishop, "The State Works of Pennsylvania," pp. 266–67; *Lancaster Intelligencer,* 9 August 1825.

41. Bishop, "The State Works of Pennsylvania," p. 265; Shelling, "Philadelphia and the Agitation in 1825 for the Pennsylvania Canal," pp. 192–97.

42. Ibid., p. 196.

43. M. Carey, *Brief View of the System of Internal Improvements of the State of Pennsylvania* (Philadelphia, 1831), p. 14.

44. *United States Gazette,* 23 September 1825.

45. *Wyoming Herald,* 2 March 1827.

46. Luzerne County Memorials and Petitions to the Pennsylvania legislature to construct a canal up the Susquehanna, House File, 51st session–1, 1826–1827, RG 7, PHMC; *Wyoming Herald,* 6 January, 1 December 1826, 26 January, 23 February, 2, 9, 23 March 1827; *United States Gazette,* 9 February 1827.

47. *Harrisburg Chronicle,* 28 February, 3, 6, 17, 24 March 1828.

48. Ibid., 4 February 1828.

49. Bishop, "The State Works of Pennsylvania," pp. 282, 284, 286.

Chapter 8

1. (JC), Table showing the relative heights and distances of several proposed routes to connect the City of New York with the Susquehanna River and Luzerne Coal Formation, JCC, WHGS.

2. George MacCulloch (hereafter cited as GM) to JC, Morristown, 6 February 1825, JCC, ANSP.

3. *United States Gazette,* 15 September, 4, 11 October, 15 November, 2, 6, 7, 12, 14, 20, 30 December 1825; Julius Rubin, "Canal or Railroad? Imitation and Innovation to the Erie Canal in Philadelphia, Baltimore, and Boston," *Transactions of the American Philosophical Society* n.s. 51, pt. 7 (1961), pp. 15–62; L. T. C. Rolt, *The Railway Revolution: George and Robert Stephenson* (New York, 1962), pp. 59–87.

4. Rubin, "Canal or Railroad," pp. 15–62; Richard Shelling, "Philadelphia and the Agitation in 1825 for the Pennsylvania Canal," pp. 175–204.

5. *Facts and Arguments in Favor of Adopting Railways in Preference to Canals, in the State of Pennsylvania* (Philadelphia, 1825).

6. GM to JC, Morristown, 20 September 1825, JCC, ANSP.

7. Ibid.

8. GM to JC, Morristown, 20 October 1825, JCC, ANSP.

9. JC, Proposal for a Coal company in the Wyoming Valley with a railroad to New York City, 28 October 1825, JCC, ANSP.

10. Ibid.

11. JC, Map of Part of the Luzerne Coal Formation Showing the Waters, Owners of Lands, Turnpikes, State, and a few of the Common Roads (1825), JCC, WHGS.

12. Sarah Cist to GMH, Wilkes-Barre, 16 December 1825, JCC, WHGS.

13. Diary of GMH, GMH Section, HC, WHGS.

14. H. M. Drinker to GMH, Covington, Luzerne County, 5 January 1826, GMH Section, HC, WHGS; W. Stroud to GMH, Stroudsburg, 22 January 1826, ibid.

15. Julius M. Bogen, *The Anthracite Railroads: A Study in American Railroad Enterprise* (New York, 1927), pp. 76–95; W. David Lewis, "The Early History of the Lackawanna Iron and Coal Company: A Study in Technological Adaptation," *Pennsylvania Magazine of History and Biography* 96 (October 1972): 424–68.

Conclusion

1. Studies in social mobility have supported earlier work on the background of early nineteenth-century entrepreneurs. Francis W. Gregory and Irene D. Neu, "The American Industrial Elite in the 1870's," in *Men in Business: Essays on the Historical Role of the Entrepreneur,* ed. William Miller (1952; reprint, New York, 1962), pp. 193–211; Lewis E. Atherton, *The Pioneer Merchant in Mid-America* (1939; reprint, New York, 1969), pp. 18–19; Edward Pessen, *Riches, Class, and Power Before the Civil War* (Lexington, Mass., 1973), pp. 92–129; Edward Pessen, ed., *Three Centuries of Social Mobility in America* (Lexington, Mass., 1974), pp. 59–121; John N. Ingham, "Rags to Riches: The Effect of City Size and Related Factors on the Recruitment of Business Leaders," *Journal of American History* 63 (December 1976): 615–37. For the pioneers of the anthracite trade see (The Delaware and Hudson Company), *A Century of Progress: History of the Delaware and Hudson Company 1828–1923* (Albany, 1925), pp. 8–62; Richard Richardson, *Memoir of Josiah White* (Philadelphia, 1873); C. K. Yearley, Jr., *Enterprise and Anthracite: Economics and Democracy in Schuylkill County, 1820–1875* (Baltimore, 1961), pp. 57–59. Yearley's statement that small businessmen dominated production in Schuylkill County requires more careful scrutiny in light of the unpublished research I have done on them during the past decade. The same point applies to Rowland Berthoff, "The Social Order of the Anthracite Region," *Pennsylvania Magazine of History and Biography* 84 (July 1965): 261–91. Concerning education for business see Thomas Cochran, *Business in American Life: A History* (New York, 1972), pp. 31–35. The best general study of merchants who lived during this period is

Glenn Porter and Harold C. Livesay, *Merchants and Manufacturers: Studies in the Changing Structure of Nineteenth-Century Marketing* (Baltimore, 1971).

2. Jacob E. Cooke, "Tench Coxe, American Economist: The Limitation of Economic Thought in the Early Nationalist Era," *Pennsylvania History* 42 (October 1975): 267–89.

3. Thomas C. Cochran, "The Entrepreneur in Economic Change," *Explorations in Economic History*, 2d ser. 3 (Fall 1965): 25–38.

4. W. W. Rostow, *The Stages of Economic Growth: A Non-Communist Manifesto* (Cambridge, 1962), pp. 1–35.

5. Pennsylvania, Department of Mines and Mineral Industries, *Anthracite Division, 1959 Annual Report* (Harrisburg, 1960), p. 8; Sydney A. Hale, ed., *The Coal Trade: The Year Book of the Coal and Coke Industry 1921* (New York, 1922), p. 108.

6. Compiled from United States Bureau of Census, *Census for the United States 1790–1970; History of Luzerne, Lackawanna and Wyoming Counties, Pa.*, (New York, 1880), p. 378; *Smull's Legislative Handbook* (Harrisburg, 1901), p. 263.

7. Alfred D. Chandler, Jr., "Anthracite Coal and the Beginnings of the Industrial Revolution in the United States," *Business History Review* 46 (Summer 1972): 141–81.

8. Frederick E. Saward, ed., *The Coal Trade* (1873–1900) (New York, 1874–1901), passim.

Index